DETECTABILITY of SPREAD-SPECTRUM SIGNALS

The Artech House Radar Library

David K. Barton, *Series Editor*

Modern Radar System Analysis by David K. Barton

Introduction to Electronic Warfare by D. Curtis Schleher

High Resolution Radar by Donald R. Wehner

Electronic Intelligence: The Interception of Radar Signals by R.G. Wiley

RGCALC: Radar Range Detection Software and User's Manual by John E. Fielding and Gary D. Reynolds

Over-The-Horizon Radar by A.A. Kolosov, *et al.*

Principles and Applications of Millimeter-Wave Radar, Charles E. Brown and Nicholas C. Currie, eds.

Multiple-Target Tracking with Radar Applications by Samuel S. Blackman

Radar Propagation at Low Altitudes by M.L. Meeks

Radar Cross Section by Eugene F. Knott, *et al.*

Radar Anti-Jamming Techniques by M.V. Maksimov, *et al.*

Radar System Design and Analysis by S.A. Hovanessian

Monopulse Principles and Techniques by Samuel M. Sherman

Monopulse Radar by A.I. Leonov and K.I. Fomichev

High Resolution Radar Imaging by Dean L. Mensa

Radar Detection by J.V. DiFranco and W.L. Rubin

Handbook of Radar Measurement by David K. Barton and Harold R. Ward

Radar Technology, Eli Brookner, ed.

Radar Range-Performance Analysis by Lamont V. Blake

Techniques of Radar Reflectivity Measurement by Nicolas C. Currie

Radar Reflectivity of Land Sea by Maurice W. Long

Aspects of Modern Radar, by Eli Brookner, *et al.*

Analog Automatic Control Loops in Radar and EW by Richard S. Hughes

VCCALC: Vertical Coverage Plotting Software and User's Manual by John E. Fielding and Gary D. Reynolds

Electronic Homing Systems by M.V. Maksimov and G.I. Gorgonov

Principles of Modern Radar Systems by Michel H. Carpentier

Secondary Surveillance Radar by Michael C. Stevens

DETECTABILITY of SPREAD-SPECTRUM SIGNALS

Robin A. Dillard
George M. Dillard
Naval Oceans Systems Center

Artech House

Library of Congress Cataloging-in-Publication Data
Dillard, Robin A., 1936-
Detectability of spread spectrum signals.
Includes bibliographies and index.
1. Signal detection. 2. Spread-spectrum communications. I. Dillard, George M., 1931-. II. Title.
TK102.5.D455 1989 621.38′043 88-34286
ISBN 0-89006-299-4

ARTECH HOUSE, INC.
685 Canton Street
Norwood, MA 02062

International Standard Book Number: 0-89006-299-4
Library of Congress Catalog Card Number: 88-34286

10 9 8 7 6 5 4 3 2 1

Contents

Preface

We joined the Navy Electronics Laboratory (NEL), San Diego, California, in the late 1950s. During the period from 1967 to 1977, the facility was known as the Naval Electronics Laboratory Center (NELC), and in 1977 it became the Naval Ocean Systems Center (NOSC). Most of our initial work was in radar signal detection, primarily automatic detection techniques. Our first supervisor, James Caspers, stressed the need for a strong theoretical background in statistics, especially in the area of hypothesis testing. Under his guidance, we explored such areas as sequential and fixed-sample-size testing, distribution-free and nonparametric detection, constant false-alarm rate (CFAR) detection, and techniques that involve binary integration, such as the binary moving-window detector.

In thc late 1960s, we were asked by a communication group to help them assess the detectability of some candidate designs of spread-spectrum communication systems. We were immediately intrigued by the problem and searched the literature for solution methods. The most useful papers were those on radiometers and energy detection. (Some of these are described in an annotated bibliography at the end of Chapter 2.) Two people already working on the project offered valuable initial guidance; they were L. E. (Skip) Hoff, NELC, and M. P. Ristenbatt, Cooley Electronics Laboratory. They had studied the problem and recognized that trade-off analysis is necessary between detection by energy integration over the entire transmission and detection of individual pulses.

We quickly realized that radar detection analysis is mathematically analogous in many ways to spread-spectrum detection analysis. In particular, the detection probability function for integrating statistically independent pulses in white, Gaussian noise also applies to energy detection. Another technique to which we easily made the transition was binary integration in the form of a binary moving-window detector. In its original radar application, the window moves in time as the antenna scans in azimuth. In the spread-spectrum application, the moving window slides, in time, to detect pulses hopped in frequency or time. The point here is that the analysis tools that we needed came primarily from the radar literature, and we were already comfortable with them. In addition, signal interception methods involving scanning in frequency or azimuth are applicable both to radar signals and communication signals.

During that initial analysis and comparison of the detectability of spread-spectrum systems in the late 1960s, we were able to provide some fairly accurate estimates when we could make simplifying assumptions. The experience was not entirely satisfying, however, because many more questions were raised than answered. Also, we reviewed the intercept analysis results of others and observed two common kinds of mistakes. One was the misuse of the normal probability function. The other involved disregarding the effect of the integration time on the false-alarm rate. We were able to address the problem again, in the early 1970s, under the Independent Research program at NELC. The analyses were significantly extended, but still were based on normal approximations to energy detection. The Naval Electronic Systems Command supported our effort in the mid-1970s to extend the analysis by using exact calculations. Additional in-house Independent Research funds enabled interesting research later in the decade, some of which is discussed in this book.

When we were asked by Artech House to write a book on spread-spectrum techniques, we saw an opportunity to examine some aspects of the problem in greater detail and to compute with improved accuracy the detector performance and comparison curves. (Our personal computer outperforms most of the computers we previously had available.) We also had hoped to find results for optimum (likelihood-ratio) detection and correlation methods, but we found the other investigations to be too time consuming. However, we were able to find several new interesting results. For example, we addressed more fully the effect of signal splitting, extended the computations for the binary moving-window detector, and devised accurate and simple approximation methods for evaluating energy-detector performance.

We are indebted to several persons who contributed directly or indirectly to the writing of this book. First, our interest in detection theory began and grew during the years we worked for James Caspers. Without his leadership and inspiration, it is unlikely we would have developed the skills necessary for success in the field. The associations, mentioned above, with Skip Hoff and M. P. Ristenbatt were responsible for our earlier work on the detection of spread-spectrum signals. Ray Lowry, engineer at NOSC, also made significant contributions to the earlier work. Dr. Ellen Stewart Kuhns and Dr. Eugene Cooper of the NOSC Independent Research program supported the work that led to publication of the paper "Detectability of Spread Spectrum Signals" in the *IEEE Transactions on Aerospace and Electronic Systems* in 1979. James Shields, Information Security Group, NOSC, read the final draft and made several helpful suggestions for improvement. Others, too numerous to mention individually, contributed by their many discussions with us during the thirty years we have worked in the field.

Chapter 1
Introduction

Development of the first spread-spectrum systems began at least four decades ago [1,2]. The early systems were designed to provide low detectability or protection from jamming or interference. Most of the applications of spread-spectrum techniques are in the fields of radar, communication, and navigation. The uses in communication alone include antijamming, anti-interference, *low probability of intercept* (LPI), message privacy, multiple-user access, and selective addressing.

This book addresses the problem of signal design, with emphasis given to LPI signal structures for communication. The intent is to help systems designers to evaluate mathematically the detectability of candidate signal structures and to select signal parameters that minimize the probability of detection. Another use is to assist in the design of intercept systems that are effective against particular signals. Detection techniques that can be used to intercept the various types of signal structures are described. Exact formulas are given for applications where the detectability of a specific signal must be determined accurately or where the signal parameters that minimize detectability must be determined exactly. Graphs and approximate formulas are presented for situations where estimates of detectability are sufficient. Many of these graphs compare detection techniques or the detectability of different signal structures. The graphs illustrate relationships between signal parameters and covertness.

We confine the discussion and analyses to the interceptor's detection problem, beginning at his receiver, except for some discussions of antenna considerations. We consider properties of interceptor antennas only when they call for a modification in the detection analyses presented for a detection system, such as an antenna scanning in azimuth. We assume that the signals to be detected are well designed, with no spurious components or other characteristics that are easily exploited. Except where indicated, the analysis assumes that the noise is from a stationary, white, Gaussian random process.

The implementation of spread-spectrum systems for a variety of applications is discussed in the open literature. Many papers concerning implementation are available, and some of the early ones are reprinted in Dixon [2]. Dixon's textbook [3] provides a good introduction and describes the advantages of various implementations. Appendix 2 of [3] is an extensive bibliography on spread-spectrum techniques and applications. A three-volume set on spread-spectrum communication by Simon *et al.* [4] also includes a chapter on the detection of LPI signals. Torrieri [5] covers secure communication and discusses the intercept problem. Schleher [6] includes a discussion of the detectability of radar and communication LPI signals. A relatively small subset of the literature on spread-spectrum signals is dedicated to the intercept problem. Wiley [7] provides an excellent coverage of methods of intercepting radar LPI waveforms. A number of journal papers outline techniques for detecting LPI signals. We will discuss some of these papers [8–12] and use their results. Reference [10] is in a special issue on spread-spectrum communication, published by the IEEE Communications Society in May 1982. This issue also includes the paper giving the origins of spread-spectrum communication [1] and one giving the theory [13].

1.1 TOPICS COVERED

Techniques applicable to the problem of detecting spread-spectrum signals are discussed in Chapter 2. The primary technique useful against weak *pseudonoise* (PN) signals is *energy detection*. Energy detection can be accomplished in a variety of ways, most commonly by using a radiometer. Several radiometer types are described, including the Dicke radiometer and the two-receiver correlation radiometer. If the transmission involves *frequency hopping* (FH) or *time hopping* (TH), also known as "time dodging," with or without the additional use of PN signaling, other detection techniques must be considered. Often, energy detection turns out to be the best choice. Simple methods of detecting hopped signals, such as binary integration, are also discussed. In Chapter 2, we introduce shorthand notation to describe the various receiver types. The chapter ends with a list of annotated references that are arranged by topic.

Methods of analyzing and calculating detection system performance are presented in Chapter 3. When analysis based simply on energy detection is sufficient, the instructions, equations, and graphs in Chapter 3 will make the job easier. The user must first be sure that the detection situation is one in which the assumption of a simple energy-detection system is adequate. If the only spread-spectrum feature of the signal is PN, simple energy detection is usually sufficient, although a dedicated interceptor may obtain a small improvement in detectability by exploiting the spectral characteristics of the signal (see, for example, Section 2.6). Sections 3.2 and 3.3 contain the formulas and graphs for estimating vulnerability to the energy detector. Careful attention must be given to formulas for *false-alarm rate* (FAR) when comparing transmissions of different lengths because the most common error in comparisons involves misuse of the inverse relationship between FAR and integration time. The FAR must be considered when comparing systems that for any reason have different decision rates.

The reader who needs to analyze the vulnerability of a particular signal structure can begin by reading the section in Chapter 4 that discusses that structure. Each section of the chapter gives references to descriptions in Chapter 2 of the detection systems appropriate to use against that signal type. (We recommend reading the first few pages of Chapter 2 on system functions and energy detection.) References are also made in Chapter 4 to the pertinent formulas and graphs in Chapter 3 with instructions for their use. The analysis in Chapter 4 serves to estimate the detectability of any particular signal structure for several detection-system types, ranging from simple to moderately sophisticated, and to select signal parameters so as to minimize detectability.

Some sections of Chapters 2 and 3 detail minor considerations of energy-detection operation (for example, "integrate-and-dump" *versus* continuous integration), which, in total, affect the analysis results by a few decibels at most. If rough estimates of detectability are adequate, or if the estimates are for comparing systems having equal losses, these sections may be disregarded.

The instructions in the above paragraphs are for systems designers with specific detection problems. The reader who is interested in the techniques and analysis of interceptor detection is advised to read the material in order.

Signal detectability is treated throughout as a statistical problem. The reader should be acquainted with concepts such as statistical independence, conditional probability, and hypothesis testing. Several references that discuss these topics are given in the annotated bibliography at the end of Chapter 2. The reader without a statistical background may use the graphs and formulas in Chapters 3 and 4, but at the risk of misusing them.

1.2 TERMINOLOGY AND NOTATION

When discussing the detectability of the many kinds of signal structures, a convenience is to have names for the individual signals or pulses within a transmission. The terminology of signal structure used here is as follows.

(1) If the transmission consists of a number of distinct signals, all having approximately the same duration and bandwidth, these signals will be called *data symbols*. (The signals may or may not represent data symbols from a communicator's viewpoint.) A data symbol can be a simple pulsed sinusoid, or it can be a spread-spectrum signal involving FH, TH, or PN. We assume that all data symbols have the same structure type within a transmission. Bursts and radar pulses also can be treated as data symbols for the purpose of making detectability calculations.

(2) When each data symbol consists of a number of pulses (e.g., a data symbol is a spread-spectrum signal), the pulses composing these data symbols will be called *elements*. Elements usually have a time-bandwidth product of approximately unity. The term *coarse structure* refers to the shape and arrangement of the data symbols within the transmission, and the term *fine structure* or *microstructure* refers to the

shape and arrangement of the signaling elements within the data symbols. In detectability calculations, the concept of elements is useful mainly when the elements of each data symbol are separated in time or space.

In Chapters 3 and 4, the expression *pulse* or *spread-spectrum pulse* is often used instead of data symbol. This is mainly because the analysis and graphs in many cases apply both to the detection systems that detect individual data symbols and systems that detect individual elements (provided that the correct parameter values are assumed). Also, the word *pulse* may mean more to the reader who does not first become familiar with the terminology described here and used throughout Chapter 2.

Table 1.1 defines the notation used for signal parameters. For communication, T_1 usually denotes the duration of the message. For radar, it generally denotes the dwell time, which is the amount of time that the scanning antenna is radiating in the direction of the interceptor. Figure 1.1 illustrates the use of notation and terminology. The most common types of spread-spectrum techniques are described below.

PN—Pseudonoise. This is a wideband, noiselike transmission. The most common is a *direct sequence* (DS) signal, generated by using a pseudorandom sequence to phase modulate the carrier at a much greater rate than the rate of the *phase-shift-keyed* (PSK) data modulation. The generated signal is sometimes filtered before transmission to flatten the power spectrum.

FH—Frequency Hopping (Hopped). A pseudorandom process assigns each pulse or data symbol to one of a large number of frequency slots, no two pulses overlapping in time. In a "pure FH" transmission: (1) the data symbol is a simple pulse ($T_2W_2 \approx 1$), (2) the frequency slots are nonoverlapping, contiguous, and equally spaced, and (3) the pulses are contiguous in time, resulting in $b_2 = T_1/T_2$. The typical pure FH transmission uses *M-ary frequency-shift-keyed* (MFSK) modulation.

TH—Time Hopping (Hopped). A pseudorandom process assigns each pulse or data symbol to one of an integer number of time slots within a hopping interval. Each hopping interval contains exactly one pulse, resulting in a duty cycle equal to the ratio of pulse duration to hopping interval duration. In a "pure TH" transmission, the data symbol is a simple pulse ($T_2W_2 \approx 1$) and all occupy the same frequency band ($W_1 = W_2$). The slot selection process often will also involve *pulse position modulation* (PPM) as the data modulation.

Hybrid. This is a combination of two or more spread-spectrum techniques. Most common are:

FH/PN—PN pulses are frequency hopped;

FH/TH—Simple pulses are hopped in frequency and time;

TH/PN—PN pulses are time hopped;

FH/TH/PN—PN pulses are hopped in frequency and time.

Table 1.1 Definitions of Signal Parameters

Signal Structure Categories	*Notation*	*Definition*
All categories.	T_1, W_1	Duration and bandwidth of transmission
	E_1	Signal energy of transmission
Data symbols each contain a number of signaling elements; i.e., the symbol has an FH, TH, or PN structure. If the structure is PN, the value of T_3 usually can be disregarded in detectability calculations.	T_2, W_2	Duration and bandwidth of data symbol
	b_2	Number of data symbols per transmission
	T_{p2}	Duration of hop interval ($T_{p2} = T_1/b_2$)
	E_2	Signal energy of data symbol
	T_3, W_3	Duration and bandwidth of elements ($T_3W_3 \approx 1$)
	b_3	Number of elements per data symbol
	T_{p3}	Duration of hop interval ($T_{p3} = T_2/b_3$)
	E_3	Signal energy of element
Data symbols each have only one signaling element.	T_2, W_2	Duration and bandwidth of data symbol ($T_2W_2 \approx 1$)
	b_2	Number of data symbols per transmission
	T_{p2}	Duration of hop interval
	E_2	Signal energy of data symbol

Other possibilities are similar to hybrids involving PN pulses, except that the PN pulses are replaced with hopped signals; thus, the data symbols are FH, TH, or FH/TH.

Linear FM. This is a form of pulse compression also known as *chirp modulation*. The instantaneous frequency of each pulse is a linear function of time. Linear FM signals generally are used for other than covert purposes. If the slope is constant (radar applications), the pulses are highly detectable with a matched filter. When the slope is modulated for communication [14–17], treat the pulses as PN, letting the pulse bandwidth W_2 be the frequency range of the pulse. The time-bandwidth area for the interceptor is then equal to the pulse-compression ratio.

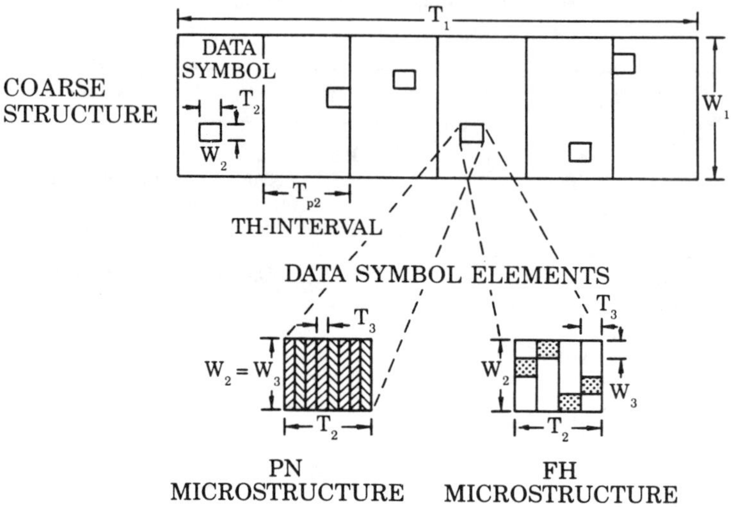

Figure 1.1 An illustration of signal parameter notation (FH/TH coarse structure and PN or FH data symbol structure).

1.3 ASSUMPTIONS ABOUT THE INTERCEPTOR

The interceptor's probability of detecting a spread-spectrum signal depends not only on the format and strength of the signal, but also on his knowledge of the signal and how dedicated he is to detecting it. We shall assume that the interceptor knows the fixed parameters of the signal and knows approximately the probability distributions of the pseudorandomly varied parameters. If the signal duration T_1 also varies (e.g., varies with the number of information bits), he has observed its distribution. He knows the signal's bandwidth W_1 and, if the signal involves hopping, he knows the bandwidth W_2 and the duration T_2 of the hopped data symbols or pulses. If data symbols are time-hopped, he knows the duration of the hop period T_{p2}. He generally cannot synchronize with the signal; that is, he must consider all possible start times for the transmission, its pulses (if hopped), and the hop period (if time-hopped). When the data symbols themselves are spread-spectrum signals, he knows the values of the fixed parameters of their microstructure (i.e., the parameters W_3, T_3, and T_{p3}).

Our basic assumption is that calculations should be based on the belief that the interceptor will dedicate his system to this signal by matching his system's parameters to the signal's parameters. (However, the equations given also apply to a "common" interceptor who attempts to detect a variety of signal types.) Unless it is unreasonable to do so, we should also assume that the interceptor will design and build the best system possible. If cost or complexity makes this assumption unreasonable, we can base the calculations on a cheaper system, such as a frequency-scanning system.

The interceptor's objective usually is to determine the presence and location of the emitter, and detecting a transmission is the first step. We address the problem of detecting weak signals, where the interceptor does not expect to determine information content. The interceptor's motive in detecting the signal is a separate issue and does not significantly affect his selection of a detection technique.

REFERENCES

1. Scholz, R. A., "The Origins of Spread-Spectrum Communications," *IEEE Trans. on Comm.*, Vol. COM-30, May 1982, pp. 822–854.
2. Dixon, R. C., ed., *Spread Spectrum Techniques*, New York: IEEE Press, 1976.. (The preface gives an historical perspective).
3. Dixon, R. C., *Spread Spectrum Systems*, 2nd Ed., New York: John Wiley and Sons, 1984.
4. Simon, M. K., *et al.*, *Spread Spectrum Communications*, Vols. I–III, Rockville, MD: Computer Science Press, 1985.
5. Torrieri, D. J., *Principles of Secure Communication Systems*, Norwood, MA: Artech House, 1985.
6. Schleher, D. C. *Introduction to Electronic Warfare*, Norwood, MA: Artech House, 1986.
7. Wiley, R. G., *Electronic Intelligence: The Interception of Radar Signals*, Norwood, MA: Artech House, 1985.
8. Urkowitz, H., "Energy Detection of Unknown Deterministic Signals," *Proc. IEEE*, Vol. 55, April 1967, pp. 523–531.
9. Dillard, R. A., "Detectability of Spread-Spectrum Signals," *IEEE Trans. on AES*, Vol. AES-15, July 1979, pp. 526–537.
10. Krasner, N. F., "Optimal Detection of Digitally Modulated Signals," *IEEE Trans. on Comm.*, Vol. COM-30, May 1982, pp. 885–895.
11. Polydoros, A., and C. L. Weber, "Optimal Detection Considerations for Low Probability of Intercept," *MILCON '82 Conf. Rec.*, Vol. 1, pp. 1–1 to 1–5.
12. Polydoros, A., and C. L. Weber, "Detection Performance Considerations for Direct-Sequence and Time-Hopping LPI Waveforms," *IEEE Trans. on Selected Areas in Comm.*, Vol. SAC-3, September 1985, pp. 727–744.
13. Pickholtz, R. L., *et. al.*, "Theory of Spread-Spectrum Communications," *IEEE Trans. on Comm.*, Vol. COM-30, May 1982, pp. 855–884.
14. Cook, C. E., "Linear FM Signal Formats for Beacon and Communication Systems," *IEEE Trans. on AES*, Vol. AES-10, No. 4, July 1974, pp. 143–150.
15. Hirt, W., and S. Pasupathy, "Continuous Phase Chirp (CPC) Signals for Binary Data Communication—Part I: Coherent Detection," *IEEE Trans. on Comm.*, Vol. COM-29, No. 6, June 1981, pp. 836–847.
16. Hirt, W., and S. Pasupathy, "Continuous Phase Chirp (CPC) Signals for Binary Data Communication—Part II: Noncoherent Detection," *IEEE Trans. on Comm.*, Vol. COM-29, No. 6, June 1981, pp. 848–858.
17. El-Khamy, S. E., and S. E. Shaaban, "Matched Chirp Modulation: Detection and Performance in Dispersive Communication Channels," *IEEE Trans. on Comm.*, Vol. 36, No. 4, April 1988, pp. 506–509.

Chapter 2
Detection Techniques

With a few exceptions presented later in this chapter, we limit the discussion of intercept systems to the detection stage that is indicated in Figure 2.1. An alarm in the detection stage activates recognition and location stages. The detection stage detects signals as optimally as practicable and, by alarming at the appropriate time, selects the data that should be processed in a recognition stage and a location stage. Signal recognition and location are not discussed further; however, recognition of signals will occur to some extent when the detection system is designed to alarm only on signals having certain characteristics.

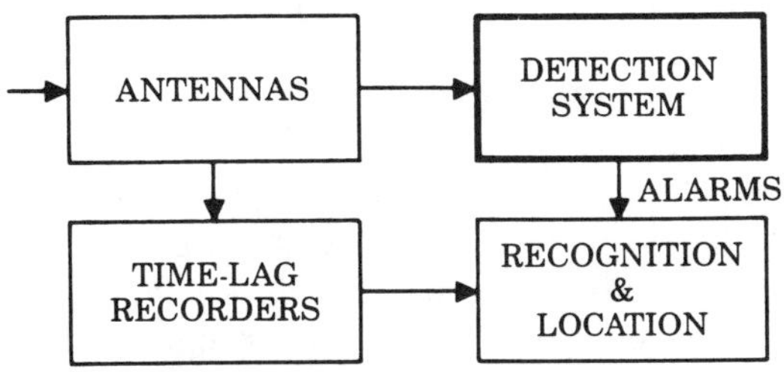

Figure 2.1 The detection system triggers the recognition and location stages

To categorize the various types of detection systems, we identify three stages, or functions. As shown in Figure 2.2, the first stage is the *predetection* stage. This stage determines the acceptance bandwidth or bandwidths of the system, and includes any frequency-scanning functions. Illustrations of detection system types will indicate the function rather than give block diagrams of the circuits required (e.g., a bandpass filter can represent both the RF and IF sections of a superheterodyne receiver). We assume that the filters have a rectangular passband, unless stated otherwise. To simplify schematics, the label "BPF (W)" represents a bandpass filter with bandwidth W.

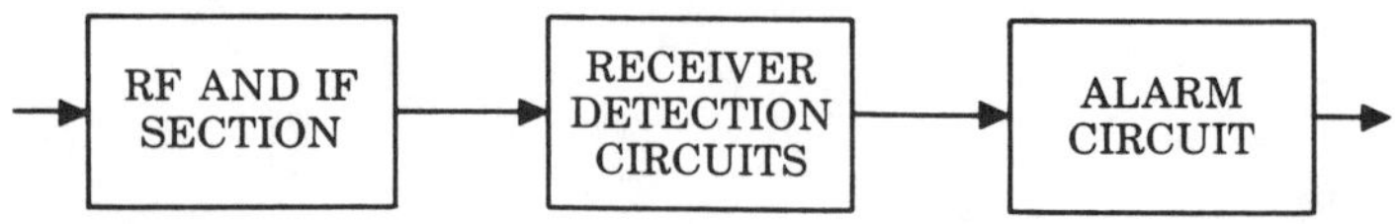

Figure 2.2 The three stages of functions used in system classification labels

The *receiver detection circuits* in Figure 2.2 perform a processing function; the function is demodulation if the interceptor knows the signal structure exactly. This stage of the system usually performs energy detection when the signals are spread in time and frequency with a low power density and the signal's microstructure is unknown to the interceptor. The *alarm circuit* can consist in practice of an operator watching a display. However, to enable detection analyses, we assume that this function is performed automatically.

The decision procedure in the alarm system uses receiver output data (either sampled or continuous) to decide whether a signal or only noise is present. The simplest procedure compares the receiver output with a threshold, and decides that a signal is present if the threshold is exceeded. This simple procedure is optimum if it is (equivalent to) the likelihood-ratio test, the optimal procedure for deciding between two (or more) hypotheses [29]. However, the likelihood-ratio test usually is much more complicated than a simple threshold comparison. The basics of the likelihood-ratio test are given in Appendix A and applications are discussed later. In references [28–36] in the annotated bibliography at the end of this chapter, the likelihood-ratio test and the related Nayman-Pearson and Bayes strategies are discussed. In particular, Root [34, p. 613] provides a good introductory discussion on the subject, and Van Trees [33] and Helstrom [32] are good sources for details on the more advanced aspects.

We indicated above that automatic alarm systems usually employ some form of thresholding. The label "S + Thr" represents a circuit that samples the input voltage, compares the sample with a selected threshold, and puts out a one if the threshold is exceeded. The comparison is analog or digital. The two possible outputs from the thresholder can be anything, and the labels one and zero are often only symbols. This operation is also termed *binary quantization*.

Some thresholding circuits do not involve sampling. For example, a *differential threshold detector* has an output of one whenever the input up-crosses a threshold. When sampling is not required (for clocked alarm-circuit functions), the label "Thr" represents the thresholding operation.

When a threshold is applied to clocked samples, a convenient way to avoid counting two or more ones per pulse is to use a *run-suppression* circuit. The circuit shown in Figure 2.3 can also reduce the effect of various types of interference. An interceptor who knows the width of the pulse can use a pulsewidth discriminator (with or without clocked sampling) to reduce interference effects. Ideally, pulses are rejected if their width is not exactly the one specified. (If the specified duration is

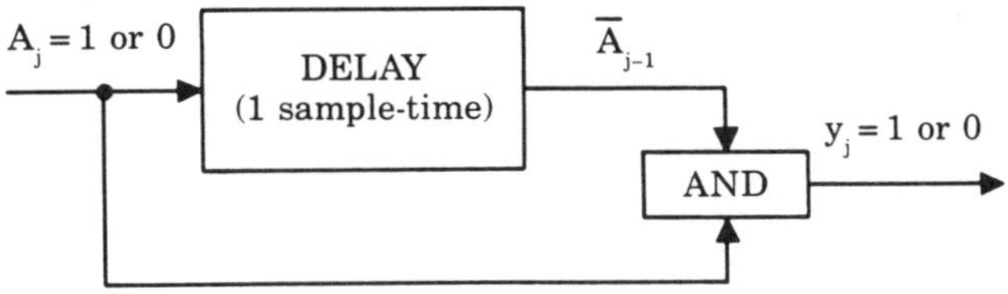

Figure 2.3 Typical run-suppressor circuit. A run of ones becomes a single one followed by zeros

approximately the reciprocal of the signal bandwidth, removal must occur before the final IF filter.)

To avoid referring to a detection system by a long description of its functions, we have devised a shorthand method for designating the names of the systems that are analyzed in Chapters 3 and 4. (These designations are summarized in Table 2.1, and are explained in Sections 2.1 and 2.2.)

Table 2.1a Summary of Detection System Designations
*S*ingle-*F*ilter *E*nergy-*D*etection Systems

Designation	*Alarm Circuit*
F-ED-T	Thresholding
F-ED-C/T	*C*ollapsing/*T*hresholding (Combine periodically spaced integration cells; threshold combined energy. Slide for all possible starting times.)
F-ED-BMW	*B*inary *M*oving-*W*indow detection (Alarm if K_M or more pulse detections occur within any T_M-second period.)
F-ED-OR/BMW	*OR*-gating/Binary Moving-Window detection (Assume synchronization within the time-hopping interval and use the knowledge that only one pulse occurs per interval by OR-gating over the interval.)
F-ED-OPT	*Opt*imum detection (Assume synchronization with the time-hopping interval and use likelihood-ratio detection.)

2.1 ENERGY DETECTION

Against many types of spread-spectrum signals having low power density, the interceptor with little knowledge of the exact structure of the signal cannot construct

Table 2.1b Summary of Detection System Designations
*F*ilter-*B*ank *E*nergy-*D*etection Systems

Designation	*Alarm Circuit*
FB-ED-T	*T*hresholding (A bank of F-ED-T systems.)
FB-ED-bank/ BMW	A *Bank* of *B*inary *M*oving-*W*indow detectors
FB-ED-BMW	*B*inary *M*oving-*W*indow detection (Use binary data from all thresholders and alarm if K_M or more pulse detections occur within any T_M-second period.)
FB-ED-OR/BMW	*OR*-gating/*B*inary *M*oving-*W*indow detection (Use knowledge that only one frequency slot is occupied at any instant by allowing at most a single 1 to enter the BMW each integration time.)
FB-ED-OR/C/ BMW	*OR*-gating/*C*ollapsing/*B*inary *M*oving-*W*indow (Each integration time, input a 1 to the C/BMW unit if at least one energy-detector output exceeds the threshold. Sum binary data from periodically spaced intervals, and threshold by comparing consecutive sums against K_M. Consecutive sums represent consecutive slot positions.)
FB-ED-OPT	*Opt*imum detection (Likelihood-ratio detection.)

a receiver that is significantly better than one that appropriately uses energy detection. Energy detection and a number of ways to employ it are discussed in this section.

2.1.1 Ideal Energy Detection

Ideal energy detection is described by the equation:

$$V = \int_{t-T}^{t} y^2(r)\, dr \tag{2.1}$$

where $y(t)$ is the input voltage, V is the output, and T is the integration time, as shown in Figure 2.4. When making detection calculations in Chapter 3 for energy detectors, we assume that the noise at the input to the energy detector is a zero-mean, stationary, Gaussian random process and it has a flat, band-limited power

Table 2.1c Summary of Detection System Designations
Frequency-*S*canning *E*nergy-*D*etection Systems

Designation	*Alarm Circuit*
S-ED-T	*T*hresholding (Set integration time to approximately the time it takes to sweep over a signal.)
S-ED-BMW	*B*inary *M*oving-*W*indow detection (Alarm if K_M or more pulse detections occur within a T_M-second interval.)
S/PC-ED-T	*T*hresholding (Threshold the integrated energy between two frequencies in the sweep path, after using *P*ulse *C*ompression.)
S/PC-ED-BMW	*B*inary *M*oving-*W*indow detection (Alarm if K_M or more pulse detections occur within a T_M-second interval. Pulses are detected by using *P*ulse *C*ompression, energy integration, and thresholding, as with the S/PC-ED-T system.)

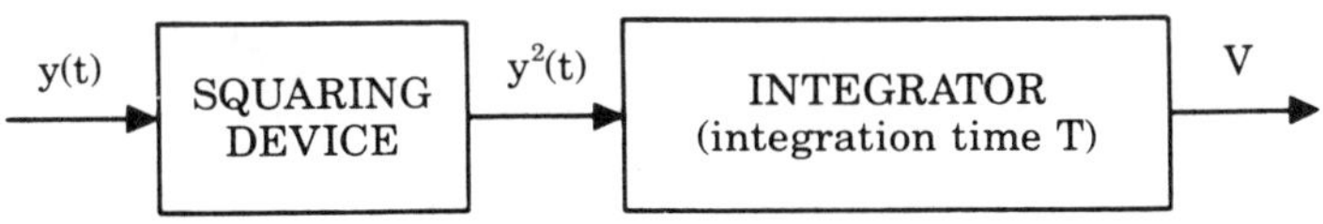

Figure 2.4 An ideal energy detector

spectral density. The normalized test statistic $V' = 2V/N_{01}$ is used to describe the output from an energy detector, where N_{01} denotes the one-sided noise power spectral density [32]. Interceptor techniques that reduce the effects of noise and interference are discussed later in this chapter, and calculation adjustments are discussed in Section 3.3.1 for situations where our assumptions about the noise do not hold.

There are two kinds of ideal energy detection. In the case of *continuous integration,* the output from the integrator at any time t is the energy integrated over the time interval $(t - T, t)$. Continuous integration ensures that the integration interval coincides with the signal time interval at some instant. This benefit is partially offset by the higher decision rate if threshold up-crossings result in false alarms. Digital integration can provide a constant weighting with time when continuous integration is desired, but usually with a small degradation [25].

In the *integrate-and-dump* (I + D) case, the output of each integrator is sampled every T seconds and then reset to zero. If two integrators having integration time T are used for each filter, then a signal or noise decision can be made every $T/2$ seconds for each filter, as illustrated in Figure 2.5. When the interceptor does

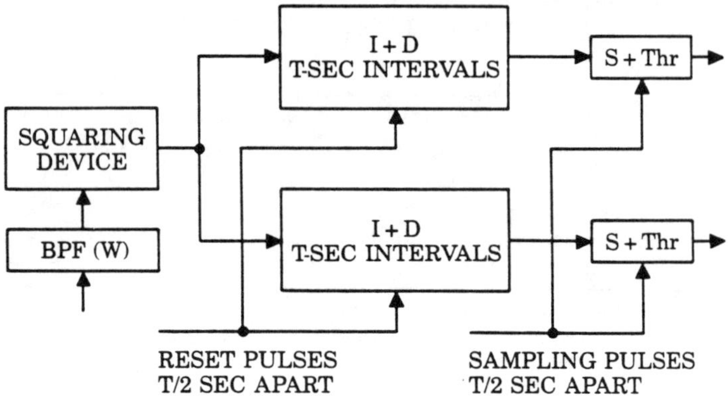

Figure 2.5 Overlapping integrate-and-dump (I+D) energy detectors

not know the exact timing of spread-spectrum pulses, but knows their duration, he can achieve improvement over conventional I + D detection by using the overlapping I + D system with T matched to the pulse duration. For detection systems designed to alarm when an energy detector output exceeds a threshold, a continuous integrating detector followed by a thresholder that responds to up-crossings has slightly better detection than an overlapping I + D detector pair (Figure 2.5) with thresholding. Additional material on this subject appears in Chapter 3.

2.1.2 Energy-Detection Receivers (Radiometers)

A receiver, or instrument, that performs an energy measurement similar to ideal energy detection is often called a radiometer if the measured radiation lies in the microwave through infrared portion of the spectrum. Descriptions of several types of radiometer receivers appear below, followed by a discussion of energy detection at lower frequencies. All of the radiometer types and other energy detection devices are indicated by ED (T), which symbolizes an energy detector with integration time T. The symbol also represents a square-law detector when the integration time is approximately equal to the reciprocal of the bandwidth. (Furthermore, square-law and linear detectors have the same theoretical detection performance when immediately followed by a thresholder [35, p. 282].)

A few of the basic types of radiometers are described here. An annotated bibliography on the subject appears at the end of this chapter. A list of typical radiometer constants (degradation factors) is given in Chapter 3 (Table 3.1), where their use in computations is explained. Computations of interceptor detection capability based on ideal energy-detector performance can easily be adjusted to apply to any assumed type of radiometer.

(a) Total-Power Radiometer. A simple receiver that performs the functions of amplifying the signal, squaring, and integrating is known as a *total-power* radiometer. It measures the total noise power received at the antenna, plus the receiver noise power. Figure 2.6 shows one configuration. If building a perfectly stable receiver were possible, the total-power radiometer would behave as the ideal energy detector described above. Because of its sensitivity to gain fluctuations, it is unsuitable for high sensitivity operation. Other radiometers described below are designed to avoid the effects of gain fluctuations.

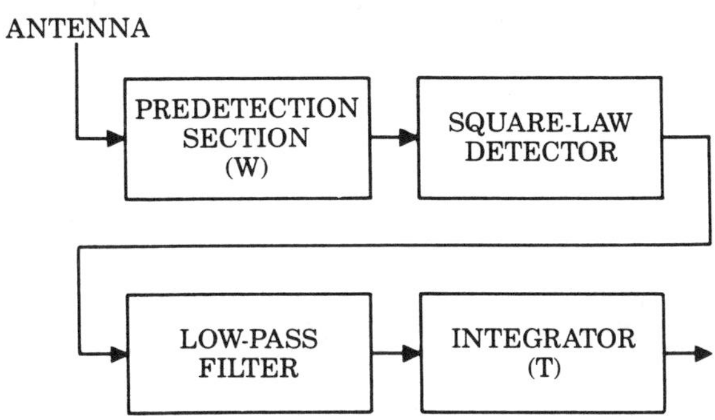

Figure 2.6 A total-power receiver

(b) Dicke Radiometer. A Dicke receiver (named for its inventor, R. H. Dicke) modulates or switches the incoming signal between the antenna terminals and a calibrated noise source. Figure 2.7 shows a typical configuration of the Dicke receiver. The product of the detected output from the receiver and a reference waveform at the same frequency as the initial modulation is filtered to produce a dc voltage when a signal is present. The switching or modulation rate is chosen to be high enough that the gain does not change within one cycle, thus eliminating the fluctuations caused by receiver instabilities. The version introduced by Dicke [1] used a mechanical modulator that continuously switched the input between the antenna and a calibrated noise source. Other waveforms now commonly used are square-wave or sine-wave modulation with sine-wave multiplication. Waveforms such as the sawtooth are also used.

(c) Two-Receiver Correlation Radiometer. The configuration of a two-receiver correlation radiometer is shown in Figure 2.8. The two inputs are multiplied together and the product is smoothed with a *low-pass filter* (LPF). This system is also known as a correlation radiometer, a two-receiver radiometer, or a product radiometer. Its principle of operation is based on the assumption that noise originating in the receivers is uncorrelated. In most cases, two antennas are required to prevent the noise at the input to one receiver from reaching the other.

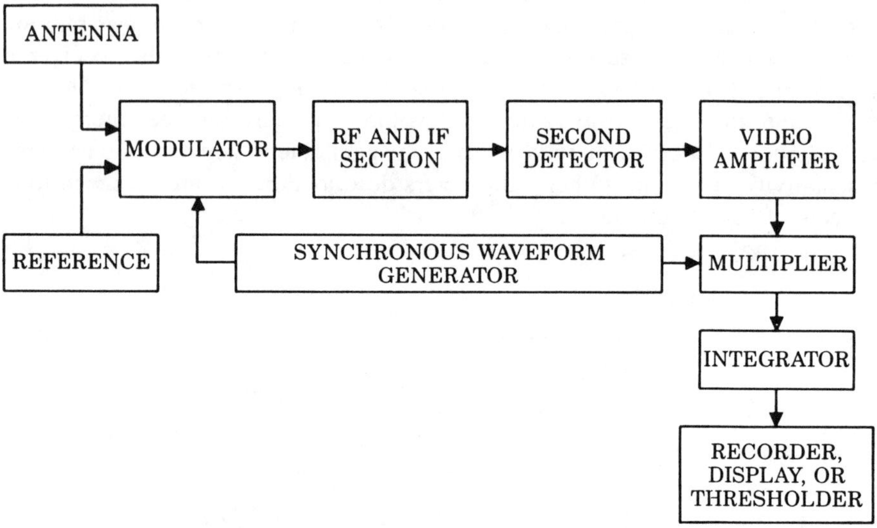

Figure 2.7 The Dicke radiometer

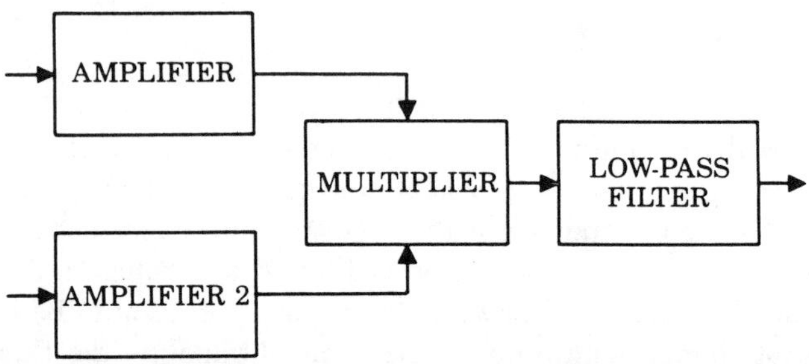

Figure 2.8 A two-receiver correlation radiometer in its simplest configuration

2.2 SINGLE-FILTER ENERGY-DETECTION SYSTEMS

Each of the detection systems described here uses energy detection at some stage of the signal processing. The energy detector can be anything from a conventional square-law detector to a circuit that integrates over a time-bandwidth area limited only by the current technology.

2.2.1 Simple Thresholding Systems

Figure 2.9 illustrates a system that integrates energy from a single frequency band (bandwidth W) and alarms whenever energy integrated over the previous T seconds exceeds the threshold. The energy detector uses either integrate-and-dump (I+D) detection or continuously integrating detection, with the thresholding circuit designed accordingly. The label F-ED-T is based on our classification scheme (see Table 2.1), the F representing a single filter. A modification of this system using two overlapping energy integrators (see Section 2.1.1 and Figure 2.5) is also included under the label F-ED-T because its detection capabilities are generally between those of the simple I+D system and the continuously integrating system.

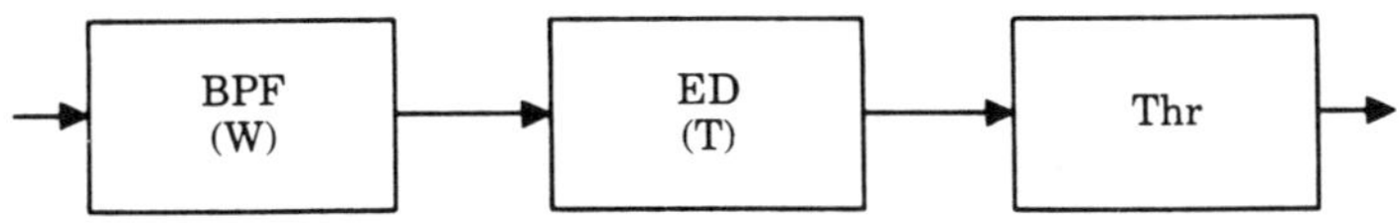

Figure 2.9 The F-ED-T system (bandpass filter, energy detector, and thresholder)

To estimate an upper bound to the detection capability of an F-ED-T system, assume that the frequency band of the detection system matches that of the transmission. If the transmitted signal band is frequency hopped from transmission to transmission, a filter-bank system or possibly a frequency-scanning detection system will be more effective.

Figure 2.10 illustrates several detection situations in which the interceptor knows the frequency band of the transmission and matches his acceptance band to the transmission band. The time-bandwidth area of integrated energy is shown at a time of peak output (possible when using continuous integration). In Figure 2.10 (a, b, and c), the interceptor integrates over the entire transmission. In Figure 2.10 (d and e), the interceptor matches his integration time to the duration of the individual pulses or data symbols (e.g., PN pulses) of the transmission. The F-ED-T system used this way is performing *simple pulse detection*. The determination of whether to integrate over the entire transmission or over each pulse usually requires calculation of detection probability for both methods. (Some graphical results of comparisons are given in Chapters 3 and 4.)

It is sometimes more realistic to estimate vulnerability to a detection system not dedicated to nor optimized for the particular signal. This may be, for example, because the interceptor lacks knowledge of the signal parameters, or his integration time is limited by antenna scanning or some other physical constraint. Many of the

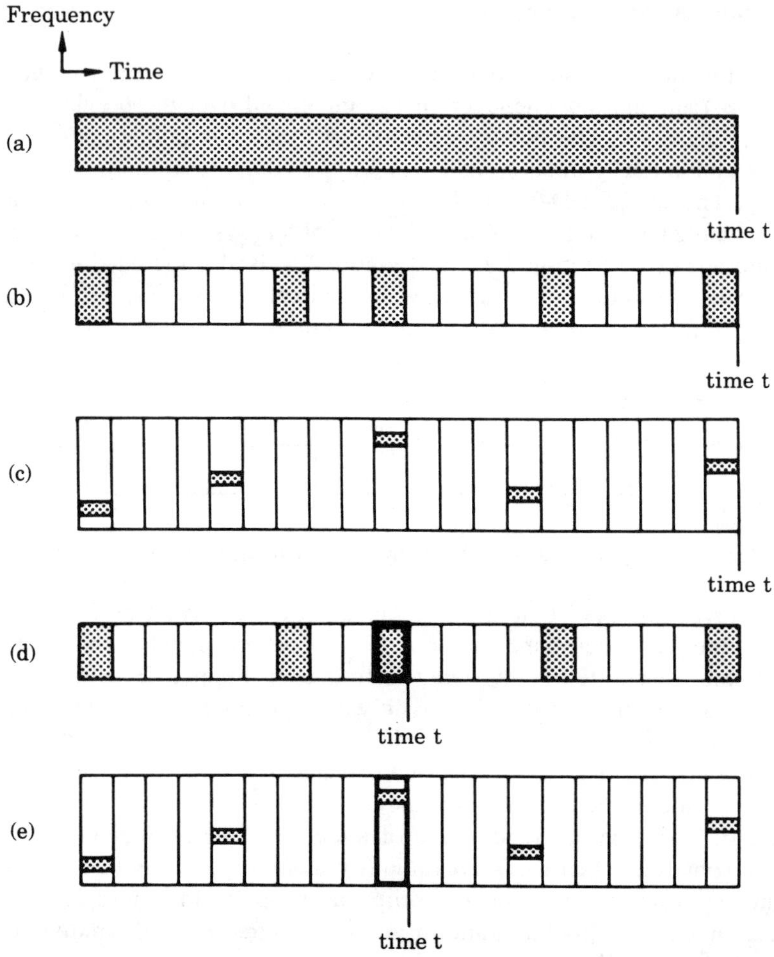

Figure 2.10 Examples of energy detection coverage for an interceptor employing an F-ED-T system

detectability calculations involve the probability that a portion of the signal coincides in time, frequency, and antenna-scan direction with a cell of energy detection. (Instructions for calculating detectability for some of these situations are given in Chapter 3.)

Table 2.2 lists the relationships between the F-ED-T detection system parameters and the signal parameters for a number of signal-structure categories. The detector threshold parameter is not included in Table 2.2 because it is a function of the false-alarm rate, as described in Chapter 3. (The notation for signal parameters was defined and illustrated in Table 1.1 and Figure 1.1.)

Table 2.2 F-ED-T (Single-*F*ilter *E*nergy-*D*etection with *T*hresholding)

Signal Categories	*Interceptor's Knowledge*	*Parameter Relationships*
All categories	Frequency band (with bandwidth W_1) and T_1	$W = W_1$; $T = T_1$ See Fig. 2.10(a-*c*)
TH coarse structure. (See note below.)	Frequency band (with bandwidth $W_1 = W_2$) and T_2	$W = W_1 = W_2$; $T = T_2$ See Fig. 2.10(d).
FH/TH coarse structure, with W_1/W_2 not large. (See note below.)	Frequency band (with bandwidth W_1) and T_1	$W = W_1$; $T = T_2$ See Fig. 2.10(e).

Note: The system F-ED-BMW discussed later has equal or greater effectiveness against TH categories, but requires relatively difficult detectability calculations. Chapter 3 gives calculation procedures for both systems.

2.2.2 Time Collapsing

The interceptor can improve his energy detection capability by *time collapsing* if he has knowledge of the time positioning of the individual signals in a transmission. For example, time collapsing is possible if the pulses or other signals in a radar transmission are equally spaced in time and the repetition rate is known. The output from an energy detector with integration time equal to the pulse duration can be inserted into a tapped delay-line integrator, with taps spaced at intervals equal to the pulse repetition period. When the periodic signal is a brief communication, the total delay of the tapped delay line ought to be equal to the message duration. For detecting a radar signal, it is to be matched to the duration of the interval during which the radar is scanning in the direction of the interceptor (i.e., the beam dwell time). Figure 2.11 is a block diagram of this detection system. The illustrated system and any that perform the same function are labeled F-ED-C/T. Systems that use cover signal information to time collapse nonperiodic signals can also be included under this label.

Table 2.3 gives the signal structure categories vulnerable to the F-ED-C/T system and, for each, the relationship between signal parameters and detection system parameters. Section 3.4.6 gives procedures for computing detection probabilities for systems employing time collapsing.

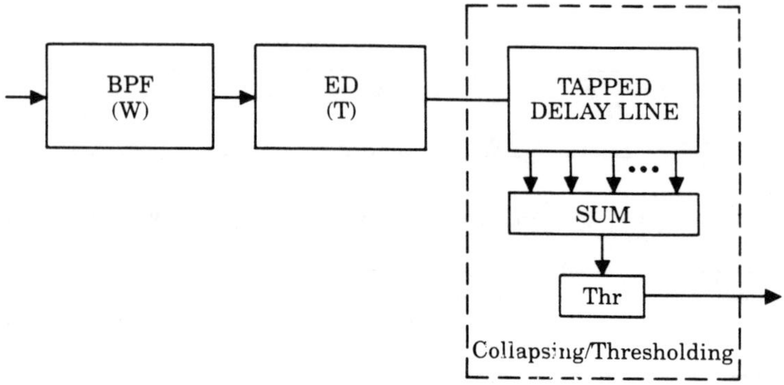

Figure 2.11 The F-ED-C/T system

Table 2.3 F-ED-C/T (Single-*F*ilter *E*nergy-*D*etection with *C*ollapsing and *T*hresholding)

Signal Categories	*Interceptor's Knowledge*	*Parameter Relationships*
PN or linear FM radar pulses periodically spaced. FH (frequency of radar pulses varied pulse-to-pulse) with W_1/W_2 large and with periodically spaced pulses.	Frequency band (with bandwidth W_1), T_1 (the beam dwell time), T_2, and T_{p2} (pulse repetition period).	$W = W_1$; $T = T_2$; Tap spacing $= T_{p2}$; Delay-line length $= T_1$. T may be greater than T_2 if pulses are jittered.

2.2.3 Binary Moving-Window Detection

For any particular type of transmission, it is not always apparent to the system designer which type of detection system is the greatest threat. For many signal-structure categories, the designer must estimate whether the interceptor will do better to integrate over the entire transmission, or to integrate separately over each of the pulses or signals composing the transmission (recall Figure 2.10). With the latter integration method the problem is to determine what further processing the interceptor can reasonably employ in his *alarm circuit* to pose the greatest threat. The F-ED-T detection system discussed earlier, with integration time matched to the pulse length, is the simplest such alarm circuit. Computations under some restrictive assumptions show it to be relatively effective. We discuss next a decision procedure that is usually more effective than the F-ED-T system against transmissions containing distinct pulses.

The expression "k-out-of-b detection" describes a rule that requires at least k-out-of-b pulses to be detected for a *signal present* decision to be made. In its simple forms, k-out-of-b detection is a common statistical decision procedure and is treated in the literature under the names binary integration; coincidence detection; binomial, fixed sample-size testing; and double-threshold detection. In most of the applications described in the literature, the properties of the binomial distribution are used to determine performance. However, in the applications of interest here, the binomial distribution does not always apply because of extra noise contributions. This is because false alarms stemming from noise in otherwise empty time-bandwidth cells can cause a condition in which there may be more than b detections.

When the message arrival time is unknown and the pulses are distributed in time, the k-out-of-b rule should be implemented with a *binary moving-window detector* (BMWD) [37]. Figure 2.12 is a simplified diagram of a detection system that uses a BMWD, labeled F-ED-BMW. The energy detector would most likely be of the I+D type. Every T seconds the output of the sample-and-threshold unit is a 0 or a 1. The run-suppressor unit shown in Figure 2.12 is needed if the I+D interval is not synchronized with pulse times. The binary integration time is T_M, usually the message duration or the radar (antenna) dwell time. At each time jT, the BMWD computes the test statistic:

$$S_{\mathrm{j}} = \sum_{k=\mathrm{j}-M+1}^{\mathrm{j}} y_k \tag{2.2}$$

and compares it with a threshold K_M. The window length is M, where $M = T_M/T$. To compute and threshold the statistic S_{j}, the BMWD, for example, could employ a digital delay-line, counter, and comparator [37]. An alarm occurs when the test statistic S_{j} crosses the threshold K_M in an upward direction; in effect, when $S_{\mathrm{j}-1} < K_M$ and $S_{\mathrm{j}} = K_M$.

The optional run-suppressor circuit in Figure 2.12 would, for example, limit the response of nonsynchronous I+D detection or overlapping I+D detection to a single 1 per pulse. A typical circuit was illustrated in Figure 2.3. Table 2.4 gives the signal structure categories against which the F-ED-BMW system is effective, and the relationships between signal parameters and detector parameters.

When the pulses or data symbols are contiguous in time (i.e., they are not time hopped or periodic), this type of detection is simply binary integration over the entire transmission, analogous to the F-ED-T system with analog integration. Binary integration has a performance advantage in this and similar cases only when strong impulse noise makes large contributions in analog integration.

If synchronization with the hopping intervals of a TH or TH/PN transmission were possible, the interceptor would modify the F-ED-BMW system shown in Figure 2.12 to use the knowledge that one and only one pulse can occur in each hopping

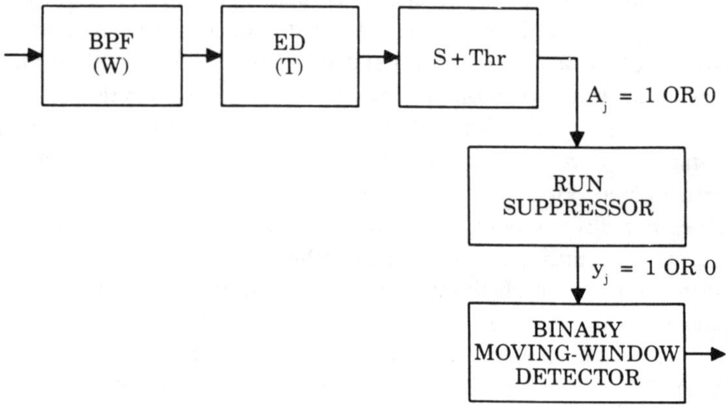

Figure 2.12 The F-ED-BMW system

Table 2.4 F-ED-BMW (Single-*F*ilter *E*nergy-*D*etection with *B*inary *M*oving-*W*indow Detection)

Signal Categories	*Interceptor's Knowledge*	*Parameter Relationships*
TH coarse structure	Frequency band (with bandwidth $W_1 = W_2$), T_1, and T_2.	$W = W_1 = W_2$; $T = T_2$; $T_M = T_1$.
FH/TH coarse structure, with W_1/W_2 not large.	Frequency band (with bandwidth W_1), T_1, T_2.	$W = W_1$; $T = T_2$; $T_M = T_1$

Note: Against TH or TH/PN signals, if the interceptor can synchronize with the hop interval, he can use the F-ED-OR/BMW system described at the end of Sec. 2.2.3 or the F-ED-OPT system described in Sec. 2.2.4.

interval. The OR function would be applied (e.g., with a binary-tapped delay line and an OR-gate) to the binary data from each hopping interval, and be followed with a BMWD having a much shorter window length. The output of the OR-gate (every T_{p2} seconds) would then be

$$Z_k = \begin{cases} 0, \text{ if } y_j = y_{j-1} = \ldots = y_{j-r+1} = 0 \\ 1, \text{ otherwise} \end{cases} \tag{2.3}$$

where r is the ratio of the hopping period to the pulse duration (i.e., $r = T_{p2}/T_2$) and $k = j/r$. The BMWD would form the sum:

$$S_k = \sum_{m=k-b_2+1}^{k} Z_m \tag{2.4}$$

We give this synchronized system the label F-ED-OR/BMW. With a reinterpretation of parameters, the detectability computations for this case are identical to those for the FB-ED-OR/BMW system to be discussed later, and many of the graphical results in Chapter 3 apply to it.

The system we have described requires synchronization with the hopping interval, but we assume that the interceptor does not know the hop interval timing. (The interceptor could stagger r BMWDs and be assured of synchronization, within a pulse duration, at some point, but the penalty is a higher FAR.) One of the main reasons for considering the synchronized case is to find an upper bound to the detection capability of pulse-detection systems against a TH or TH/PN transmission. To define the true upper bound, we examine next the form of a likelihood-ratio detector under the assumption of synchronization.

2.2.4 Likelihood-Ratio Detection

To define an upper bound to the performance of a pulse-detection system against TH and TH/PN transmissions, we assume that the interceptor can align his operations on the signal-hopping interval. This and our other assumptions are the same as those for the F-ED-OR/BMW system described above. The hopping period is T_{p2}, the energy integration time is the pulse duration ($T = T_2$), and their ratio is $r = T_{p2}/T_2$. Again, we know the interceptor can synchronize only by chance, but we wish to find the upper bound to his detection capability.

We label as F-ED-OPT the single-filter energy-detection system having its decision logic based on a likelihood ratio, the optimum detection procedure. (Likelihood-ratio detection is discussed in Appendix A, where derivations of likelihood ratios for various signal types are given.) After each hopping interval, the F-ED-OPT system performs a summing operation over the r most recent measurements $(V_{j-r+1}, \ldots, V_{j-1}, V_j)$ of the energy detector output. For detecting a pure TH transmission (with $T_2W_2 \approx 1$), the sum after j measurements is

$$S_j = \sum_{m=j-r+1}^{j} I_0(2\sqrt{V_m E_2}/N_{01}) \tag{2.5}$$

where I_0 is the modified Bessel function of the first kind of order zero, E_2 is the pulse energy, and N_{01} is the one-sided noise power spectral density. Against TH/PN signals, S_j is replaced by (see Appendix A):

$$S_j = \sum_{m=j-r+1}^{j} \frac{I_{TW-1}(2\sqrt{V_m E_2/N_{01}})}{(2V_m)^{(TW-1)/2}} \tag{2.6}$$

where $TW = T_2W_2$ and $I_n(x)$ is the modified Bessel function of the first kind of order n. The F-ED-OPT system then computes the product of the b_2 most recent sums (a sum for each hopping interval in the transmission) to form the likelihood ratio:

$$L = \prod_{m=k-b_2+1}^{k} S_m \tag{2.7}$$

and compare it with a threshold. If the logarithm of both sides is taken, the operation becomes a sum rather than a product. Unlike any other detection system considered in this chapter, the optimum pulse-detection receiver requires an estimate of the signal energy.

2.3 FILTER-BANK ENERGY-DETECTION SYSTEMS

Our primary interest in filter-bank energy-detection systems (also known as *channelized radiometer* systems) is their effectiveness against FH transmissions. Another situation addressed is when the frequency band of a communication message varies from transmission to transmission. (Detection calculations for both situations are given in Chapter 3.) The detection of FH transmissions is discussed next.

Frequency hopping usually occurs as part of the coarse structure of a transmission, the hopping of data symbols. Another use of FH to consider is where the elements, simple pulses within the data symbols, are hopped in frequency. For convenience, the hopped signals, whether data symbols, elements of data symbols, or radar pulses, often are referred to here as "pulses."

The interceptor knows that the pulses will lie within a particular band of frequencies of bandwidth W_1, and he uses a bank of bandpass filters to cover that frequency range. To make the analysis in Chapter 3 tractable, we assume ideal rectangular filters. The filter bandwidth W is matched to the pulsewidth. If the filters are nonoverlapping, the total number of filters is W_1/W. If the filters overlap by one-half (if the center frequencies are separated by $W/2$), the number of filters is $2W_1/W - 1$. (In Chapter 3, we discuss the effect on detectability of using overlapping or nonoverlapping filters.)

To simplify computations, we divide the time-frequency space currently being examined into time-bandwidth energy-integration cells by using a grid formed by the filter bands and the energy integration times. This grid is illustrated in Figure 2.13 for an FH coarse structure and an FH fine structure. When the microstructure is FH, the interceptor can alternatively use a much finer grid, one that matches the cells to the smaller pulses of size T_3W_3. Whether we assume the filters are I+D, overlapping I+D, or continuous is not important. We account for these factors in calculations by making simple parameter adjustments.

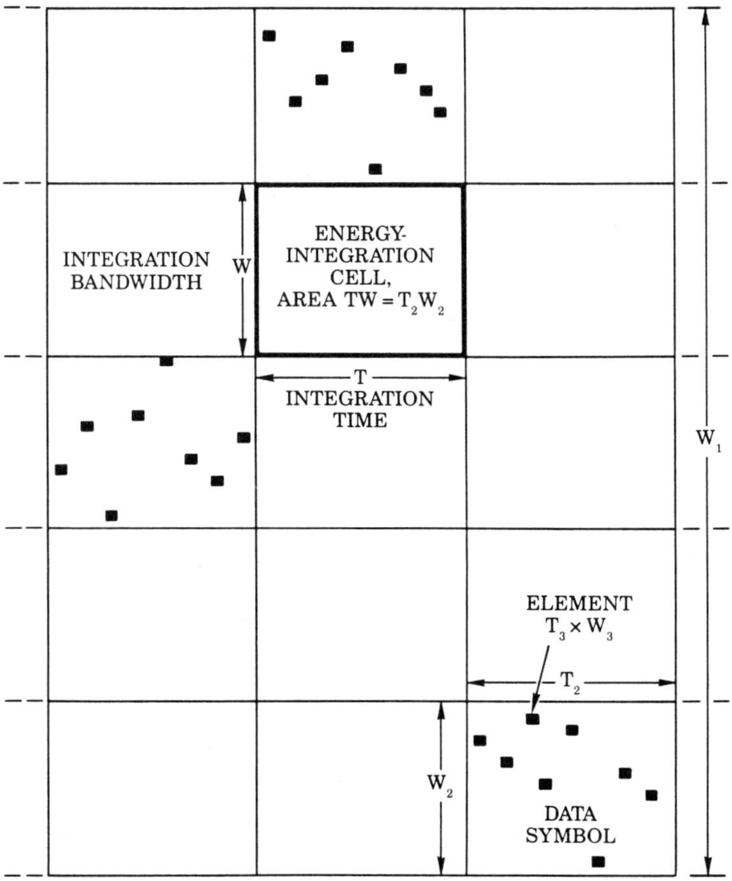

Figure 2.13 A grid dividing the time-frequency space under surveillance into energy-integration cells (shown aligned with data symbol structure)

2.3.1 Simple Channelized Systems

The simplest filter-bank system is the one illustrated in Figure 2.14, labeled FB-ED-T. Each filter is followed by an energy detector and a thresholder. Therefore, the system consists of a bank of F-ED-T systems. The discussions in Section 2.1 on I+D *versus* continuous integration and thresholding circuits also apply here. Table 2.5 lists signal categories and the associated parameters against which the system is effective. Generally, the system is used as a simple pulse-detection system.

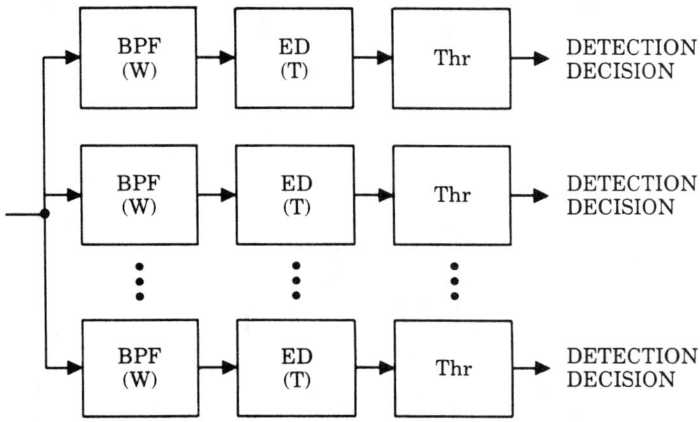

Figure 2.14 The FB-ED-T system

2.3.2 Channelized Binary Moving-Window Detection

The FB-ED-bank/BMW system is a bank of F-ED-BMW systems (Figure 2.12). The word "bank" is included in the label to distinguish the system from others below. Table 2.6 gives the signal categories the FB-ED-bank/BMW system can be used against.

An FB-ED-BMW system performs a moving-window integration of all the binary data, one bit per filter per energy-integration interval. This is equivalent to summing the outputs of the binary moving-window integrators in a FB-ED-bank/BMW system and performing the thresholding operation on that sum. This system is appropriate for use against signal structures where pulses occur at more than one frequency simultaneously.

The FB-ED-OR/BMW system illustrated in Figure 2.15 uses the knowledge that only one frequency slot is occupied at any instant by allowing at most a single 1 to enter the moving-window detector each T seconds (assuming I+D). The run-suppressor units shown in Figure 2.15 are needed if the I+D interval is not synchronized with the pulse times. The BMWD sums the 0 and 1 "decisions" over the

Table 2.5 FB-ED-T (*F*ilter-*B*ank *E*nergy-*D*etection with *T*hresholding)

Signal Categories	*Interceptor's Knowledge*	*Parameter Relationships*
FH or FH/TH coarse structure.	Frequency band (with bandwidth W_1), W_2, T_2.	Filter-bank covers signal band with W_1/W_2 nonoverlapping filters or up to $2W_1/W_2$-1 overlapping filters each having bandwidth $W = W_2$; $T = T_2$.
FH or TH coarse structure with FH fine structure (elements of data symbols are hopped).	Frequency band (with bandwidth W_1), W_3, T_3.	Filter-bank band covers signal band with W_1/W_3 nonoverlapping filters or up to $2W_1/W_3$-1 overlapping filters with each filter having bandwidth $W = W_3$; $T = T_3$.
Categories in Table 2.2 for F-ED-T except transmission band may vary from message to message.	Same, respectively, as in Table 2.2 except frequency band of next transmission is unknown.	Same as for F-ED-T (for respective category) except filter-bank band covers message-hopping band.

Note: The effectiveness of the FB-ED-OR/BMW system (discussed in Sec. 2.3.2) against the first two categories is equal to or greater than that of the FB-ED-T; however, the system requires additional interceptor knowledge and its detectability calculations are more difficult.

previous M energy integration intervals and thresholds the sum. The OR-gate function has additional value in the presence of impulse noise. Table 2.7 lists the vulnerable signal structures and the parameter relationships.

When the carrier frequency is hopped on a pulse-to-pulse basis and the pulses are periodically spaced, a more appropriate detection system is the FB-ED-OR/C/BMW system illustrated in Figure 2.16. Table 2.8 gives the signal categories and parameter relationships for the FB-ED-OR/C/BMW system.

Figure 2.16 illustrates the principle of operation of the FB-ED-OR/C/BMW system. A similar configuration for a simple binary moving-window detector is described in [37] and [38]. The method described there is applicable to all of the binary moving-window systems described in Chapter 2. In practice, the system would more likely be implemented by using a digital memory instead of the indicated shift register. By appropriately varying the memory address, data from intervals separated by T_{p2} seconds are accessed and summed.

Table 2.6 FB-ED-bank/BMW (*F*ilter-*B*ank *E*nergy-*D*etection with *B*inary *M*oving-*W*indow Detection: A Bank of F-ED-BMWs)

Signal Categories	*Interceptor's Knowledge*	*Parameter Relationships*
FH coarse structure with TH fine structure (elements of data symbols hopped in time).	Frequency band (with bandwidth W_1), W_2, T_2, T_3.	Filter-bank covers signal band with W_1/W_2 nonoverlapping filters or up to $2W_1/W_2$-1 overlapping filters each having bandwidth $W = W_2$; $T = T_3$; $T_M = T_2$.
Categories in Table 2.4 for F-ED-BMW, except transmission band may vary from message to message.	Same, respectively, as in Table 2.4 except the frequency band of next transmission is unknown.	Same as for F-ED-BMW (for respective category) except filter-bank band covers message-hopping band.

Note: The interceptor having sufficient information about the signal can use two stages of binary moving-window detection; the FB-ED-bank/BMW system followed by the alarm circuit of the FB-ED-OR/BMW (also discussed in this section) would be effective but difficult to analyze.

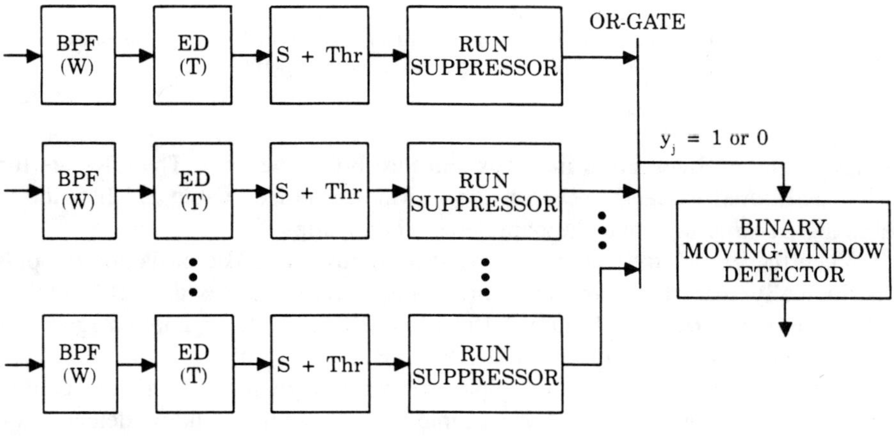

Figure 2.15 The FB-ED-OR/BMW system

Table 2.7 FB-ED-OR/BMW (*F*ilter-*B*ank *E*nergy-*D*etection with *OR*-Gating and *B*inary *M*oving-*W*indow Detection)

Signal Categories	*Interceptor's Knowledge*	*Parameter Relationships*
FH or FH/TH coarse structure. See note 1.	Frequency band (with bandwidth W_1), T_1, W_2, T_2.	Filter-bank band covers signal band with W_1/W_2 nonoverlapping filters or up to $2W_1/W_2$-1 overlapping filters each of bandwidth $W = W_2$; $T = T_2$; $T_M = T_1$.
FH or TH coarse structure with FH fine structure (elements of data symbols are hopped). See note 2.	Frequency band (with bandwidth W_1), T_1, W_3, T_3.	Filter-bank band covers signal band with W_1/W_3 nonoverlapping filters or up to $2W_1/W_3$-1 overlapping filters each of bandwidth $W = W_3$; $T_M = T_1$.

Notes:

1. If the interceptor can synchronize with the time-hopping interval, he also can employ an OR-gate over the hop interval, as described in Sec. 2.2.3 for the F-ED-OR/BMW system. For a system having N filters, the BMW would accept at most a single 1 from each block of NT_{p2}/T_2 time-frequency cells. Alternatively, he could employ the FH/TH or FH/TH/PN version of the F-ED-OPT system described in Sec. 2.2.4.
2. If $W_1 >> W_2$ or $T_1 >> T_2$, the dedicated interceptor could use two stages of BMW detection; however, analysis would be very difficult.

2.3.3 Channelized Likelihood-Ratio Detection

A single-filter system (F-ED-OPT) employing a likelihood-ratio alarm circuit was described in Section 2.2.4 for TH and TH/PN signals. The alarm circuit for a filter-bank system (FB-ED-OPT) is similar, but the likelihood-ratio is based on the joint probability distribution of the outputs of all filters. Variations of the system are tailored to specific categories of signal structure with fixed parameters. Appendix A gives derivations of likelihood ratios for various signal types. We first describe a version of the FB-ED-OPT system designed for detecting pure FH transmissions.

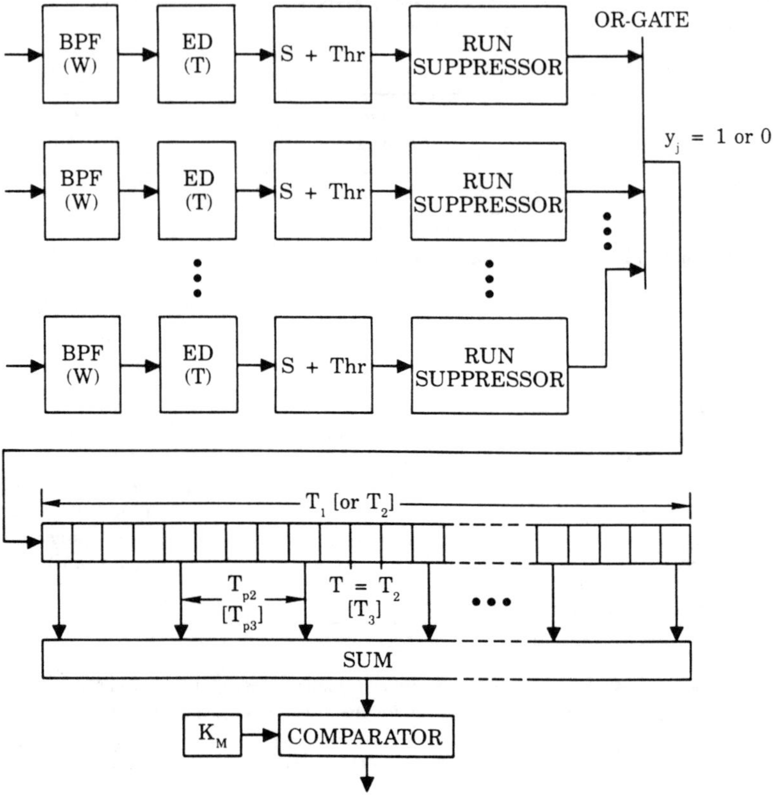

Figure 2.16 The FB-ED-OR/C/BMW system. Bracketed parameters apply to data symbols consisting of periodic elements

Recall that a pure FH signal consists of simple sinusoidal pulses that are contiguous in time. The likelihood-ratio system for detecting pure FH signals is labeled FB-ED-OPT(FH) and is shown in Figure 2.17. The filter bandwidth W equals the pulse bandwidth W_2 and the integration time T equals the pulse duration T_2, with $T_2W_2 \approx 1$. N_{01} is the one-sided noise power spectral density. (The system can also be used against data symbols having a pure FH structure, by using the parameter values $W = W_3$, $T = T_3$, $b = b_3$, and $E = E_3$.)

Against FH/PN signals, the likelihood-ratio operation shown in Figure 2.17 is replaced with the following, where $TW = T_2W_2$:

$$\prod_{m=j-b_2+1}^{j} \sum_{i=1}^{N} \frac{I_{TW-1}\left(2\sqrt{V_{im}E_2/N_{01}}\right)}{(2V_{im})^{(TW-1)/2}} \tag{2.8}$$

Table 2.8 FB-ED-OR/C/BMW (*F*ilter-*B*ank *E*nergy-*D*etection with *OR*-Gating, *C*ollapsing, and *B*inary *M*oving-*W*indow Detection)

Signal Categories	*Interceptor's Knowledge*	*Parameter Relationships*
Periodically spaced pulses vary in frequency from pulse to pulse.	Frequency band (with bandwidth W_1), T_1 (the beam dwell time), T_2, W_2, and T_{p2} (the pulse repetition period).	Filter-bank band covers signal band with W_1/W_2 nonoverlapping filters or up to $2W_1/W_2$-1 overlapping filters each with bandwidth $W = W_2$;1 $T = T_2$, tap spacing $= T_{p2}$; delay-line length $= T_1$; window length $M = b_2$.

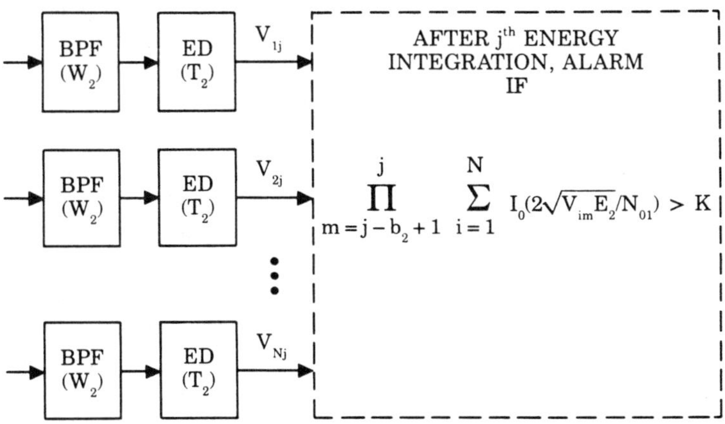

Figure 2.17 The FB-ED-OPT (FH) system for use against pure FH signals ($T_2W_2 \approx 1$)

The likelihood-ratio detectors for FH/TH and FH/TH/PN signals require the unrealistic assumption of perfect alignment with the hopping interval. These detectors serve as a model for bounding detectability, as does the F-ED-OPT system discussed earlier. Like the F-ED-OPT system, the detectors perform a summing operation and a product each hopping interval. Assume the filter bandwidth is $W = W_2$ and the integration time is $T = T_2$, and let r denote the ratio T_{p2}/T_2, where T_{p2} is the hop-interval duration. Against FH/TH signals, the system forms a sum over the most recent block of Nr samples from the energy detectors:

$$S_{\mathrm{j}} = \sum_{m=\mathrm{j}-r+1}^{\mathrm{j}} \sum_{i=1}^{N} I_0\left(2\sqrt{V_{im}E_2}/N_{01}\right) \tag{2.9}$$

The system then computes the product L of the results of the summing operation over the most recent b blocks, where $b = b_2$, the number of pulses in the transmission. Thus,

$$L = \prod_{m=k-b+1}^{k} S_m \tag{2.10}$$

An alarm occurs when the likelihood ratio L exceeds a specified threshold. Against FH/TH/PN signals, replace the summand in (2.9) with the summand in (2.8).

2.4 FREQUENCY SCANNING RECEIVERS

A frequency-scanning receiver can be adapted to detect spread-spectrum pulses of low power density by using an energy detector (functionally) in place of the usual second detector and video amplifier. (When $T \approx 1/W$, the energy detector is a square-law detector, and its performance is theoretically identical to that of a linear detector, if thresholded.) The discussion of detection systems employing energy detection continues here with the description of frequency-scanning systems.

The predetection circuits of all frequency-scanning receivers are functionally symbolized by a sweep generator (or oscillator), a mixer, and a bandpass filter. The slope of the receiver sweep path (frequency/time) is s. As before, the acceptance band of bandwidth W is assumed to be rectangular unless stated otherwise. In addition, compressive receivers employ a *pulse-compression filter,* represented by PCF.

The interceptor scans the surveillance band of bandwidth W_s with a linear sawtooth sweep. Figure 2.18 is a diagram of the scan pattern. The swept oscillator bandwidth is W_0, and the receiver sweep period is $T_s = W_0/s$. We disregard oscillator flyback time.

Figure 2.19 illustrates a frequency-scanning system in which the detection of a single pulse is sufficient to cause a message-present decision. According to the classification procedure introduced at the beginning of Chapter 2 (Table 2.1), we label this detection system S-ED-T, where the S represents frequency *scanning* (or *sweeping*).

Figure 2.20 shows a frequency-scanning system, labeled S-ED-BMW, with the alarm circuit consisting of a binary moving-window detector. In this case, K_M pulse detections are required for a message-present decision. (The BMWD is described in Section 2.2.3. A run-suppressor unit should precede the BMWD if I+D detection is used.)

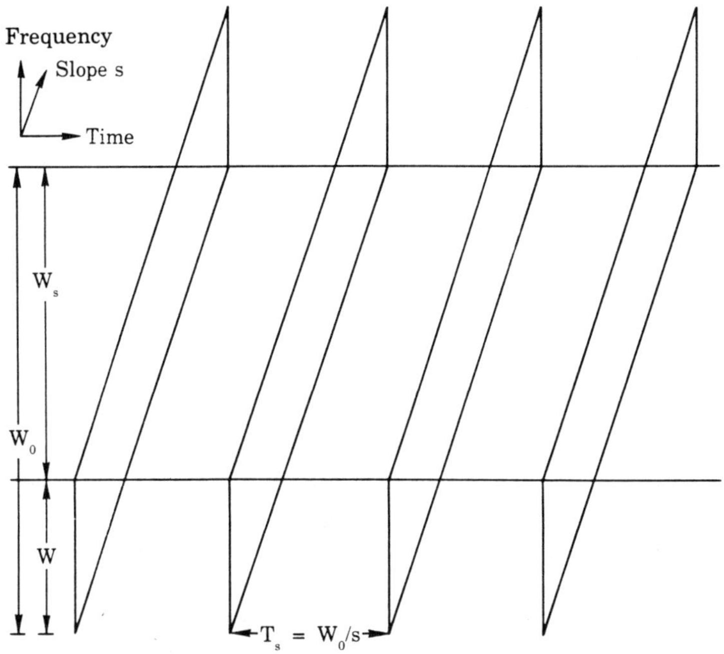

Figure 2.18 Sweep pattern for a receiver repeatedly scanning a surveillance bandwidth $W_s = W_0 - W$

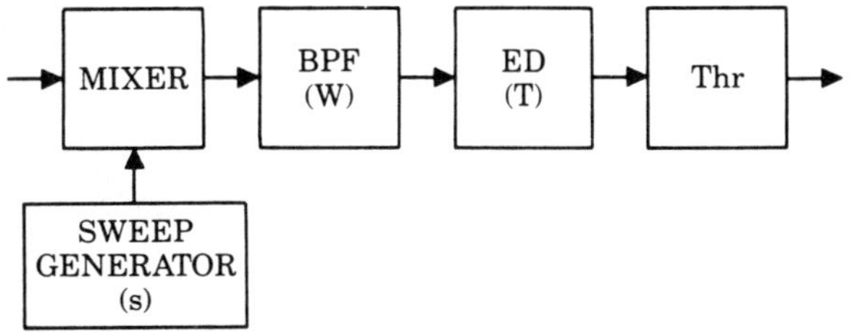

Figure 2.19 The S-ED-T system

In the situation of interest here, the interceptor's signal-to-noise ratio (SNR) will be small enough that he will have to sweep over a significant portion of a spread-spectrum pulse to detect it. The integration time should be approximately equal to the time the receiver takes to sweep over the pulse. The selection of optimum values

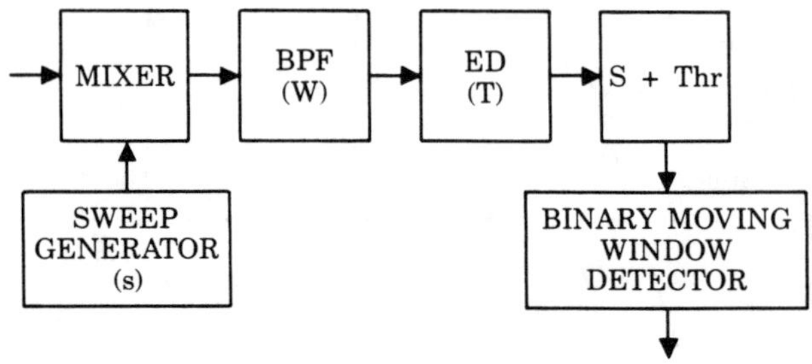

Figure 2.20 The S-ED-BMW system

of T and W is discussed for a number of signal structures in Section 3.6. Tables 2.9 and 2.10 give near-optimum values of detector parameters for the S-ED-T and S-ED-BMW systems, respectively. (The scaling ratio r in Table 2.9 is chosen to optimize the bandwidth W.) Some of the parameter relationships are better understood by first studying Figure 2.21. Tables 2.9 and 2.10 do not include signal situations in which the F-ED-T system or F-ED-C/T system will always be better. In all cases listed in these tables, a properly designed filter-bank system is better than any frequency-scanning system; however, a scanning receiver often can be more easily implemented than a filter-bank system that requires a large number of filters.

Table 2.9 S-ED-T (Frequency-*S*canning *E*nergy-*D*etection with *T*hresholding)

Signal Categories	*Interceptor's Knowledge*	*Parameter Relationships*
FH or FH/TH coarse structure.	Frequency band (with bandwidth W_1), W_2, T_2.	$W_s = W_1$; $W = r(W_2 + T_2 s)$. Choose s and r ($.5 < r < 1$) according to Sec. 3.6. If best $s > W_2/T_2$, assume an S/PC-ED-T or S/PC-ED-BMW system instead. $T = T_2$ (for $s < W_2/T_2$).
TH coarse structure and transmission band varied from message to message.	W_2, T_2. Frequency band of transmission is unknown.	Same procedure for parameter selection as above except W_s = message-hopping band.
FH or TH coarse structure with FH fine structure (elements of data symbols are hopped).	Frequency band (with bandwidth W_1), W_3, T_3	Same as above, but with T_2 and W_2 replaced with T_3 and W_3.

Table 2.10 S-ED-BMW (Frequency *S*canning *E*nergy-*D*etection with *B*inary *M*oving-*W*indow Detection)

Signal Categories	*Interceptor's Knowledge*	*Parameter Relationships*
FH or FH/TH coarse structure.	Frequency band (with bandwidth W_1), W_2, T_2.	Use the same procedure for parameter selection as given for the S-ED-T system, for the respective category. Instructions for selecting K_M are given in Secs. 3.5 and 3.6.
TH coarse structure and transmission band varied from message to message. See note.	W_2 and T_2 known. Frequency band of the transmission unknown.	Same as above.
FH or TH coarse structure (elements of data symbols are hopped).	Frequency band (with bandwidth W_1), W_3, T_3.	Same as above.

Note: Because the true pulse detections will all be in the same frequency band, a bank of BMWDs (one for each band) would perform much better than a single one, and would also identify the band.

Figure 2.22 is a functional representation of a compressive receiver with energy detection and thresholding. The system is labeled S/PC-ED-T, where PC is for *pulse compression*. The pulse compression filter (PCF) is a dispersive filter with a linear delay-*versus*-frequency characteristic matched to the sweep slope. When a BMWD is used in the alarm circuit, the resulting system, shown in Figure 2.23, is designated S/PC-ED-BMW.

The use of pulse compression in a frequency-scanning receiver is described and analyzed by Butson *et al.* [47] for continuous wave (CW) signals. The effect of a pulse compression filter, when the dispersion of the filter is matched to the sweep slope s, is to allow energy integration over frequency instead of time (along the sweep path in the original time-frequency plane). Figure 2.24 illustrates this effect. The compressive receiver is also known as a *microscan receiver*, for its ability to detect while sweeping very rapidly.

The quantity W/s is commonly known as the *differential delay* or *dispersion time* through the dispersive device, across the IF bandwidth W. The *frequency resolution* of the compressive filter is approximately equal to the inverse of the dis-

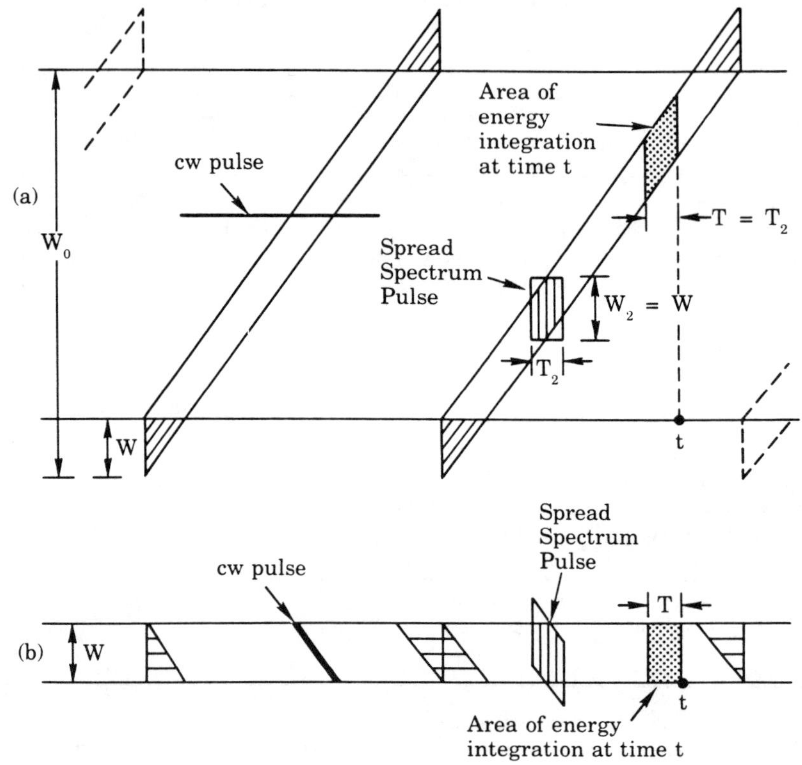

Figure 2.21 Relation between input and output for a frequency-scanning receiver without pulse compression: (a) sweep path in IF; (b) output of IF filter (input to energy detector)

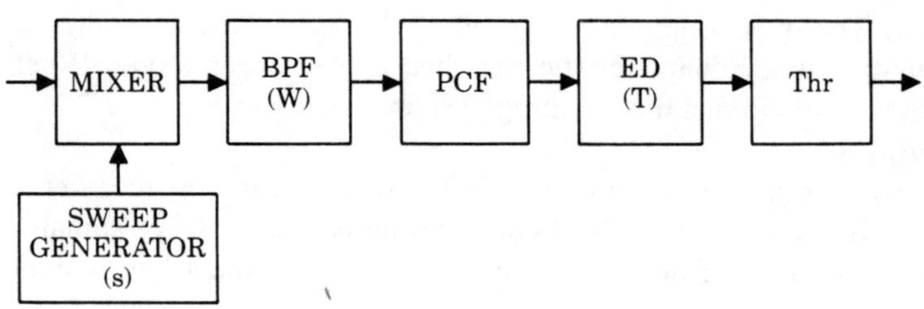

Figure 2.22 The S/PC-ED-T system

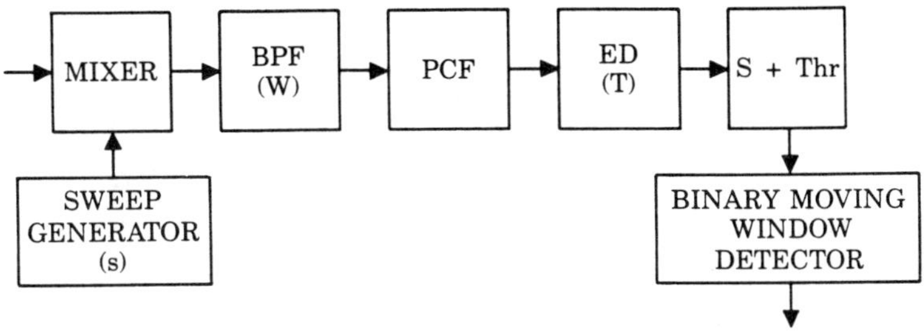

Figure 2.23 The S/PC-ED-BMW system

persion time. Frequency resolution is of interest mainly in the detection of narrow-band signals. The product of the differential delay (W_2/s) and the IF bandwidth is known as the *time-bandwidth product* or *dispersion factor* [47] of the compressive filter. When considered as the ratio of the input duration W/s to the output duration $1/W$ of a swept CW segment, it is called the *compression ratio*. When energy detection follows compression, a more useful definition of compression factor is the ratio of the differential delay to the integration time T.

Tables 2.11 and 2.12 give the relationships of detector parameters and signal parameters for the S/PC-ED-T and S/PC-ED-BMW systems, respectively. The parameter r is chosen to optimize the filter bandwidth $W = r(W_2 + T_2s)$.

2.5 DETECTION IN NONSTATIONARY, NON-GAUSSIAN ENVIRONMENTS

A mathematical analysis of energy detectors is feasible only under the assumption that the signals are embedded in stationary, white Gaussian noise. At higher frequencies, this assumption is satisfied because the primary noise source is thermal noise from within the receiver; therefore, detector performance can be satisfactorily predicted. At the lower frequencies, however, external noise predominates and energy detector performance is degraded. We discuss next the properties of external noise and methods of coping with it. We then discuss how the interceptor can use distribution-free or nonparametric testing in order to maintain a constant false-alarm rate.

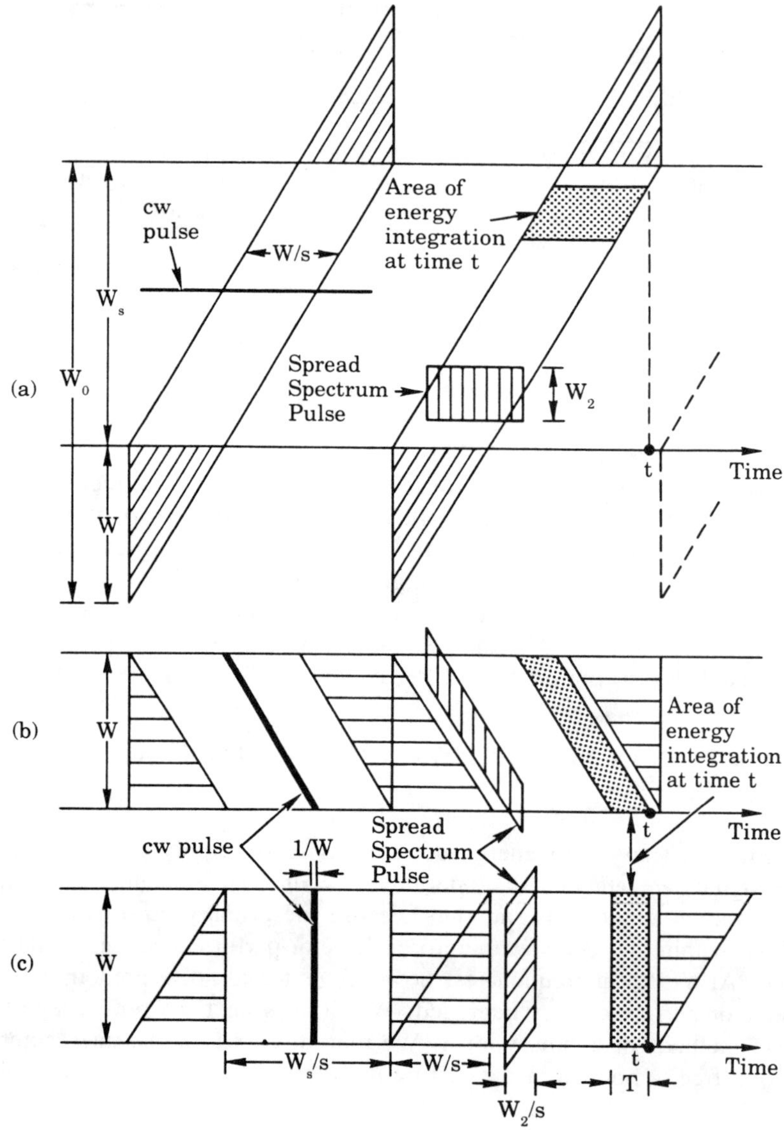

Figure 2.24 Relation between input and output for an ideal compressive receiver: (a) sweep path in RF; (b) output from IF amplifier; (c) output from compressive filter (consider the delay at the lowest frequency as zero.)

Table 2.11 S/PC-ED-T (Frequency-*S*canning with *P*ulse *C*ompression and *E*nergy-*D*etection with *T*hresholding)

Signal Categories	*Interceptor's Knowledge*	*Parameter Relationships*
FH or FH/TH coarse structure.	Frequency band (with bandwidth W_1), W_2, T_2.	$W_s = W_1$; $W = r(W_2 + T_2 s)$. Choose s and r $(.5 < r < 1)$ according to Sec. 3.6. If best $s \leq W_2/T_2$, assume an S-ED-T or S-ED-BMW system instead. $T = W_2 s$ (for $s > W_2/T_2$).
TH coarse structure and transmission band varied from message to message.	W_2 and T_2 known. Frequency band of transmission is unknown.	Same procedure as above except W_s = message-hopping band.
FH or TH coarse structure with FH fine structure (elements of data symbols are hopped).	Frequency band (with bandwidth W_1), W_3, T_3.	Same procedure as above but with W_2 and T_2 replaced with W_3 and T_3.

2.5.1 Detection at Lower Frequencies

The main frequency region of concern at "lower" frequencies is the widely used high-frequency (HF) band, covering 3–30 MHz. At these frequencies, external noise generally exceeds receiver noise, and increases inversely with frequency [54, 55]. The three dominant kinds of noise are atmospheric, extraterrestrial (galactic), and manufactured. The source of atmospheric noise is lightning charges. Atmospheric noise propagates long distances via ionospheric reflection, but is fairly constant except during local thunderstorms. The season and time of day strongly affect the level of atmospheric noise. Extraterrestrial noise is caused by the sun (e.g., solar flares), stars, and interstellar space, and becomes greater than atmospheric noise above 10 MHz [55]. The source of manufactured noise is electrical equipment and is, of course, greatest in populated areas.

Atmospheric noise spikes are a major cause of errors to the intended receiver at HF frequencies. The spikes can also result in an intolerable false-alarm rate for channelized intercept systems that alarm on a single threshold crossing. A binary moving-window system should fare much better, but will generally require a higher

Table 2.12 S/PC-ED-BMW (Frequency-*S*canning with *P*ulse *C*ompression and *E*nergy-*D*etection with *B*inary *M*oving-*W*indow Detection)

Signal Categories	*Interceptor's Knowledge*	*Parameter Relationships*
FH or FH/TH coarse structure.	Frequency band (with bandwidth W_1), W_2, T_2.	Use the same parameter selection procedure as given for the S/PC-ED-T system, for the respective category. Instructions for selecting K_M are given in Secs. 3.5 and 3.6.
TH coarse structure and transmission band varied from message to message. See note.	W_2 and T_2 are known. Frequency band of transmission is unknown.	Same as above.
FH or TH coarse structure with FH fine structure (elements of data symbols are hopped).	Frequency band (with bandwidth W_1), W_3, T_3.	Same as above.

Note: Because the true pulse detections would all be in the same frequency band, a bank of BMWDs (one for each band) could perform much better than a single one and also could identify the band.

value of the threshold K_M (the required number of pulse detections) than that optimum for white, Gaussian noise. Energy integration over the transmission or over spread-spectrum pulses can also be preceded by amplitude peak clipping, to reduce the effect of large noise spikes.

There are several other techniques the interceptor can use to reduce the effects of external noise and interference. Various anti-interference circuits can be used to discriminate against unwanted signals on the basis of amplitude or shape. Examples were given earlier for pulse-length discrimination. Another example is the common use of notch filters to selectively attenuate continuous narrow-band interference.

Multipath propagation degrades communication performance at long ranges but will not significantly affect intercept probabilities. *Signal fading* is another serious problem at HF. The amplitude of the signal fluctuates over a period of from a fraction of a second to a few minutes. Both the intended receiver and the interceptor can improve reception of fading signals by using space diversity because antennas spaced several wavelengths apart usually do not suffer fading simultaneously. The use of separated antennas with two-receiver correlation is discussed in the references under the topic of radiometers at the end of this chapter.

Our concern with energy detection at the lower frequencies is from the standpoint of estimating interceptor detection performance. Section 3.3.1 discusses a method of adjusting energy-detector performance to make these estimates.

2.5.2 Distribution-Free and Nonparametric Methods

When the noise or signal-pulse-noise distributions are unknown, conventional procedures can produce poor results. If a decision procedure is based on wrong assumptions about the distributions, the detection performance can be nonsensical. Distribution-free methods of signal detection require a minimal number of assumptions and can be used in many cases when the underlying signal and noise distributions are unknown. In general, a test for deciding between signal-plus-noise and noise is said to be distribution-free if its false-alarm probability remains essentially constant as long as the noise distribution belongs to a particular, generally large, class of distributions. The term *nonparametric* is often used interchangeably with *distribution-free,* although they are not synonymous, and there are various definitions of nonparametric. Another term used to describe a special case of distribution-free procedures, especially in the literature on radar, is *constant false-alarm rate* (CFAR) procedures.

Distribution-free and nonparametric signal detection procedures are discussed in references [40–46]. Reference [44] analyzes two easily implemented distribution-free systems designed to detect communication signals, and finds there is little loss in detectability compared with optimum procedures that require knowledge of the probability distribution of the observed data.

2.6 DETECTION TECHNIQUES OTHER THAN ENERGY DETECTION

Detection and recognition techniques can also be based on spectral and autocorrelation analysis and on the exploitation of statistical properties of signals. Most of these techniques require higher SNR for initial detection than do the energy-detection techniques described in Section 2.1. One technique that is useful against biphase PN sequences is the "multiplication phase-lock receiver," which can detect and recognize biphase or certain other conventional signals. When a biphase signal is frequency doubled, the second harmonic of the (suppressed) carrier is obtained. If a coherent component is detected at the second harmonic but not at the fundamental, the presence of a biphase signal is indicated.

2.7 ANNOTATED REFERENCES

This section contains a set of annotated references on the subjects discussed in Chapter 2. The references are divided into topics, but are numbered consecutively. Under each topic the listing is chronological. The topics are:

Radiometers;

Signal Detection Theory;

Binary Moving-Window Detection;

Distribution-Free and Nonparametric Testing;

Compressive Receivers;

High-Frequency (HF) Channels.

Not all of the references listed here are cited in the text.

Radiometers

1. Dicke, R. H., "The Measurement of Thermal Radiation at Microwave Frequencies," *Review of Scientific Instruments*, Vol. 17, July 1946, pp. 268–275.
Dicke discusses the connection between Johnson noise and blackbody radiation, by using a simple thermodynamic model. He describes a microwave radiometer and its theory of operation. He also describes a method of calibrating by using a variable temperature resistive load. (The radiometer Dicke described here became widely used and was named for him. The implementation he described used square-wave modulation and multiplication. Other authors have considered sine-wave variations.)

2. Goldstein, S. J., Jr. "A Comparison of Two Radiometer Circuits," *Proc. IRE,* Vol. 43, No. 2, Pt. 1, November 1955, pp. 1663–1666.
This paper compares the Dicke circuit, for both sinusoidal and square-wave modulation, with a two-receiver radiometer. Goldstein's derivations here of the minimum detectable signal power are discussed in later issues of *Proc. IRE.*

3. Green, J. C., "Stability Requirements and Calibration of Radiometers when Measuring Small Noise Powers," *Proc. IRE,* Vol. 45, No. 3, March 1957, pp. 359–360.
This note discusses the stability requirements for radiometers when measuring small noise temperatures and proposes a modification of the Dicke system to alleviate the severe stability requirements. The modification involves a variable noise source coupled to one arm of a directional coupler that is inserted between the antenna terminals.

4. Tucker, D. G., M. H. Graham, and S. J. Goldstein, Jr., "A Comparison of Two Radiometer Circuits," *Proc. IRE,* Vol. 45, No. 3, March 1957, pp. 365–366.

This correspondence consists of three letters concerning a paper by S. J. Goldstein, Junior (reference [2], above). The "Letter from Mr. Tucker" discusses a number of points left out of Goldstein's paper. The "Letter from Mr. Graham" points out that the formula for least detectable signal power for the two-receiver radiometer should contain a factor of 2, for a realistic comparison with other radiometers. The "Author's Comment" by Goldstein discusses some minor errors in his earlier paper.

5. Strom, Leland D., "The Theoretical Sensitivity of the Dicke Radiometer," *Proc. IRE,* Vol. 45, No. 9, September 1957, pp. 1291–1292.
The results of a "new" analysis are reported here along with comments on the remarks of Goldstein and Tucker (reference [4], above). Strom found Goldstein's revisions in error, and stated that the original result of Goldstein (reference [2], above) was correct, although it was based upon a slightly incorrect autocorrelation function. A correction to Strom's paper appears in *Proc. IRE,* December 1957. J. Galejs finds Strom's comments on Goldstein's 1957 note to be incorrect.

6. Galejs, J., "Comparison of Subtraction-Type and Multiplier-Type Radiometers," *Proc. IRE,* Vol. 45, No. 10, October 1957, pp. 1420–1422.
This letter compares the thresholds of subtraction-type and multiplication-type radiometers and establishes rates of amplifier gain variations that result in nonessential (3 dB or less) threshold degradation of the subtraction-type radiometer. A graph of SNR *versus* γ for constant observation time T, where γ is the low-pass filter bandwidth, illustrates that the SNR is maximized if $\gamma = 0.2/T$. Another figure gives the least detectable signal power and the permissible output drift rate as a function of T for a square-law detector radiometer. If the output drift rate does not exceed the limit specified in this figure, the subtraction-type radiometer is comparable in sensitivity with others. The author concludes that the subtraction-type radiometers appear to merit consideration if the integration time is on the order of seconds or less.

7. Strum, P. D., "Considerations in High-Sensitivity Microwave Radiometry," *Proc. IRE,* Vol. 46, No. 1, January 1958, pp. 43–53.
This paper discusses considerations in high-sensitivity microwave radiometry that the author thinks to be prominent but have been ignored in previous analyses. He shows that the predetection bandwidth should be maximized for highest sensitivity and discusses the improved techniques available to permit increased bandwidth. A table of relative threshold powers is presented for variations of modulation and filtering. The table includes results for three types of detectors: an average detector, a peak-envelope detector, and a peak-peak detector. A brief correction by R. S. Colvin appears in *Proc. IRE,* Vol. 47, December 1959, p. 2105.

8. Drake, F. D., and H. I. Ewen, "A Broadband Microwave Source Comparison Radiometer for Advanced Research in Radio Astronomy," *Proc. IRE,* Vol. 46, No. 1, January 1958, pp. 53–60.
This paper discusses the successful development of a radiometer system operating

at centimeter wavelengths. The effects of gain fluctuations, serious when trying to achieve high sensitivities, are eliminated by introducing compensating noise.

9. Graham, Martin, "Radiometer Circuits," *Proc. IRE,* Vol. 46, No. 12, December 1958, p. 1966.
This short correspondence describes a radiometer with two receivers and one antenna and gives the least detectable signal figures for both square-wave and sine-wave multiplication.

10. Colvin, R. S., *A Study of Radio-Astronomy Receivers,* Ph.D. Thesis, Stanford University, October, 1962.
The author's objective is to furnish a consistent basis for predicting and comparing the performance of different systems. Explicit, general definitions for the parameters of bandwidth and integration time are given. A variety of receivers are analyzed and their performance is compared with a total-power receiver. Tables of integration time and bandwidths for common types of filters were calculated. The author discusses such topics as automatic gain control (AGC) systems, modulated pilot-signal systems, and a microwave spectroheliograph receiver.

11. O'Donnell, S. R., *A Comparison of Radiometers,* M. Sc. Thesis, Ohio State University, 1963.
Expressions are developed for the least detectable signal of various receiver types by using only the statistical properties of the receiver, receiver noise, and signal. The radiometer types analyzed are the total-power, the correlation, the Dicke, the phase-switching, and Graham's. All types are assumed to use a single-section RC integrating filter.

12. Kelly, E. J., *et al.,* "The Sensitivity of Radiometric Measurements," *J. Soc. Indust. App. Math.,* Vol. 11, No. 2, June 1963, pp. 235–257.
Background material and definitions are presented. The dc radiometer sensitivity and the effect of radiometer gain fluctuations are analyzed. The sensitivity of a general type of radiometer is evaluated, and performance is optimized over design parameters.

13. Cohn, M., *et al.,* "High-Sensitivity 100- to 300-Gc Radiometers," *Proc. IRE,* Vol. 51, No. 9, September 1963, 1227–1232.
This paper describes high-sensitivity radiometers built to operate at frequencies within the 2.1 mm and 1.3 mm atmospheric absorption windows. Their measured performance demonstrated that high-sensitivity superheterodyne-type radiometers can be built to operate at the shorter millimeter wavelengths.

14. Ohm, E. A., and W. W. Snell, "A Radiometer for a Space Communication Receiver," *Bell System Tech. Journal,* Vol. 42, No. 5, September 1963, pp. 2047–2080.
By adding a square wave of noise to the input of an ultra-low-noise receiver via a directional coupler, a radiometer with a sensitivity greater than a Dicke type can be

achieved when the basic system temperature is less than 18 K. A noise-adding radiometer is compatible with a communications receiver and has been used (1) to measure and monitor the absolute system temperature and (2) to check the boresighting of a satellite communication antenna by detecting and tracking radio stars.

15. McGillem, C. D., and T. V. Seling, "Influence of System Parameters on Airborne Microwave Radiometer Design," *IEEE Trans. on Mil. Electronics,* Vol. MIL-7, No. 4, October 1963, pp. 296–302.
The principles of operation of broadband microwave radiometers are discussed. Formulas are given for the performance of various receiver configurations, and the effects of variations in stability of gain, noise figure, and reference signal are discussed. A range equation is derived, but it applies mainly to airborne mapping radiometers. Most of the results appear in Chapter 39 of *Radar Handbook,* edited by M. I. Skolnik (McGraw-Hill, 1970). A figure gives five radiometer receiver configurations. A table gives radiometer constants for nine circuit configurations. These constants differ from the usual constants because the author includes a factor of 1/2 to account for the antenna accepting only one polarization.

16. Evans, G., "Optimum Design of Radiometer Input Stages," *Proc. IEEE,* Vol. 52, No. 9, September 1964, pp. 1057–1058.
In this note, the author shows that, for optimum radiometer sensitivity, the contribution to the effective noise temperature from the main part of the radiometer should be one-third of the contribution from the aerial and first stage.

17. Tiuri, M. E., "Radio Astronomy Receivers," *IEEE Trans. on A & P,* Vol. AP-12, December 1964, pp. 930–938.
A survey of the principles of radio astronomy receivers is presented. System noise temperature, the sensitivity of different receiver types, and the calibration of receivers are studied. Among the receivers discussed are the gain-modulated Dicke, the null-balancing Dicke, Graham's (two receivers, one antenna), the correlation, and the phase-switching (two receivers, two antennas) radiometers. The author presents tables of predetection and postdetection noise bandwidths for some common types of low-pass filters and integrating circuits and a table of radiometer constants for the various receiver types discussed. The tables and some of the textual material are contained in Chapter 39 of *Radar Handbook,* edited by M. I. Skolnik (McGraw-Hill, 1970).

18. Tiuri, M. E., "Radio-Telescope Receivers," Ch. 7 in Kraus, J. D., ed., *Radio Astronomy,* New York: McGraw-Hill, 1966. pp. 236–293.
This chapter contains much of the material given in the author's paper (reference [17] above). With some minor changes in notation and terminology, he gives block diagrams of receiver types and tables of predetection and postdetection equivalent bandwidths for various filter types. He includes additional receiver types, derivations of equations, and tutorial material.

19. Urkowitz, H., "Energy Detection of Unknown Deterministic Signals," *Proc. IEEE,* Vol. 55, No. 4, April 1967, pp. 523–531.
In this paper, Urkowitz derives the statistical distribution of the output of an energy measuring device (an ideal total-power radiometer) operating in white Gaussian noise. He gives a chi-square approximation and a normal approximation to the noncentral chi-square cumulative distribution, the distribution occurring in the signal-present case. Related work in energy detection by J. I. Marcum and E. L. Kaplan is discussed. Sets of receiver operating characteristic curves are given for several time-bandwidth products ($2TW$ = 2, 10, 20, 30, 50) and an extended nomogram of the chi-square cumulative distribution is given.

20. Robertson, G. H., "Computation of the Noncentral Chi-Square Distribution," *Bell System Tech. Journal,* Vol. 48, January 1969, pp. 201–207.
This article gives a formula that allows accurate values of the cumulative noncentral chi-square distribution to be computed. By using a computing algorithm based on this formula, detection probabilities for energy detection in white Gaussian noise can be computed for time-bandwidth products as great as 10,000.

21. King, D. D., "Passive Detection," Ch. 39 in Skolnik, M. I., ed., *Radar Handbook,* New York: McGraw-Hill, 1970, pp. 39–1 to 39–36.
This chapter starts with equations and principles that provide a basis for radiometry. Although most of the material is intended for applications such as mapping, tracking, or target identification, much of it also applies to intercept receivers. Examples of radiometric systems are given in considerable detail. Many of the results given are from a paper by Tiuri (reference [17], above) and a paper by McGillem and Seling (reference [15], above). Included are tables of bandwidth ratios for different types of IF filters and integrating filters, and a table of radiometer constants.

22. Aitken, G. J. M., and R. J. Mills, "A Radiometer Employing Frequency-Domain Coding," *Radio Science,* Vol. 5, No. 3, March 1970, pp. 535–539.
This paper describes a radiometer in which the spectrum of the noiselike signal is shaped (spectral coding) so that it can be distinguished from the receiver noise by a simple correlation process. This radiometer has a small-signal sensitivity comparable to that of the Dicke (temporally coded) system. A practical system was constructed to demonstrate the technique. A table of comparisons of radiometer sensitivities gives the sensitivity factors for three Dicke circuits and two spectral-coded circuits.

23. Ural, A., *et al.,* "Analysis of the Switched Self-Balancing Radiometer," *IEEE Trans. on AES,* Vol. AES-6, No. 2, March 1970, pp. 239–247.
Analysis of the switched self-balancing comparison radiometer with coupling between channels is given. An expression for the power spectrum, at the output of the radiometer is derived. It is shown that measurement errors due to interchannel coupling can be corrected by phase switching. Radiometer sensitivity is also calculated.

24. Hagfors, T., and J. M. Moran, "Detection and Estimation Practices in Radio and Radar Astronomy," *Proc. IEEE,* Vol. 58, No. 5, May 1970, pp. 743–759.
This paper is concerned mainly with astronomical applications, but has a brief review of radiometer types.

25. Ohlson, J. E., "Efficiency of Radiometers using Digital Integration," *Radio Science,* Vol. 6, No. 3, May 1970, pp. 341–345.
The performance of total-power and balanced Dicke-type radiometers using digital integration is considered. The square-law device output is considered to be smoothed by an RC filter and then sampled. Degradation due to finite sampling rate and quantization noise is derived. A figure gives curves of the degradation factor *versus* normalized sampling interval for T/t_{RC} = .1, .5, 1, 3, 10, and infinity, where T is the averaging block time (integration time) and t_{RC} is the filter time constant. In practice $t_{RC} < T$ is a constraint.

26. Evans, G., and C. W. McLeish, *RF Radiometer Handbook,* Norwood, MA: Artech House, 1977 (148 pp.).
This book provides guidelines for the design and use of radiometers. It gives simple derivations for most of the facts presented. When detailed mathematics is necessary, intuitively based explanations are given. Six appendices are included that provide additional material on topics such as radiometer sensitivity and bandwidth definitions.

27. Hersman, M. S., and G. A. Poe, "Sensitivity of the Total Power Radiometer with Periodic Absolute Calibration," *IEEE Trans. on MTT,* Vol. MTT-29, No. 1, January 1981, pp. 32–40.
An expression is derived for the sensitivity of a total-power radiometer. This expression accounts for noise in calibration measurements and the effects of random gain fluctuations. The analysis is verified by comparing measurements (from a 94-GHz receiver) with predictions. The performance of an example total-power radiometer system is examined, and some numerical results are given.

Signal Detection Theory

28. Van Meter, D., and D. Middleton, "Modern Statistical Approaches to Reception in Communication Theory," *IRE Trans. on Inf. Theory,* Vol. IT-4, September 1954, pp. 171–212.
This paper briefly discusses the difficulties in the signal-to-noise ratio approach and examines various statistical criteria and decision rules.

29. Peterson, W. W., *et al.,* "The Theory of Signal Detectability," *IRE Trans. on Inf. Theory,* Vol. IT-4, September 1954, pp. 171–212.
The authors show that a receiver that calculates the likelihood ratio for each receiver input is the optimum receiver for detecting signals in noise. A number of different

ensembles of signals in band-limited white Gaussian noise are considered and a possible receiver design is discussed for each case.

30. Selin, I., *Detection Theory,* Princeton, NJ: Princeton University Press, 1965. (128 pp.).
From the author's preface: "The book is not a reference book. Rather it is an exposition of a number of topics. . . . The material is framed in a rather general context, but all of the topics treated in detail deal with normal noise."

31. Hancock, J. C., and P. A. Wintz, *Signal Detection Theory,* New York: McGraw-Hill, 1966 (247 pp.).
Chapter 3 of this text introduces simple-hypothesis testing and applies the concepts to the problem of detecting signals in noise. Chapter 7 formulates optimum receiver structures for random channels.

32. Helstrom, C. W., *Statistical Theory of Signal Detection,* 2nd Ed. Oxford: Pergamon Press, 1968 (467 pp.).
The author describes his textbook as "an elementary introduction to the detection of signals in radar and communication technology." The book contains a wealth of material, but requires an extensive errata sheet, available from the author.

33. Van Trees, H. L., *Detection, Estimation, and Modulation Theory,* Part I, New York: John Wiley and Sons, 1968 (697 pp.).
Part 1 and subsequent volumes were based on notes prepared by the author for a graduate course at the Massachusetts Institute of Technology. He states that a knowledge of elementary probability theory and second-moment characterization of random process is assumed, and some familiarity with matrix theory and linear algebra is helpful.

34. Root, W. L., "An Introduction to the Theory of the Detection of Signals in Noise," *Proc. IEEE,* Vol. 58., No. 5, May 1970, pp. 610–623.
This invited, tutorial paper presents a simple treatment of some of the basic problems in detection theory and refers the reader to the appropriate sources for the more sophisticated and elaborate theory. Several pages are devoted to discussion of likelihood criteria. The bibliography contains sixty-four references.

35. Whalen, A. D., *Detection of Signals in Noise,* New York: Academic Press, 1971 (411 pp.).
From the author's preface:

> "This book is intended to serve as an introduction to the principles and applications of the statistical theory of signal detection. . . . The author hopes the book can be used for self-teaching, and serve as well as a reference book for the practicing engineer."

36. Root, W. L., "Remarks, Mostly Historical, on Signal Detection and Signal Parameter Estimation," *Proc. IEEE,* Vol. 75, No. 11, November 1987, pp. 1446–1457.

This paper contains "Historical Remarks on Signal Detection" that are divided into two categories: pre-1953 and post-1953. Included are discussions of noise theory, which established a foundation for early statistical work, and parameter estimation. Seventy-three references are given, including the early works of Schottky and Nyquist.

Binary Moving-Window Detection

37. Dillard, G. M., "A Moving-Window Detector for Binary Integration," *IEEE Trans. on Inf. Theory,* Vol. IT-13, No. 1, January 1967, pp. 2–6.
A binary moving-window detector that performs binary integration on receiver output data is described. A detailed description of its implementation in a radar application is given. A similar implementation for a communication application is discussed in reference [38], below.

38. Dillard, G. M., *Binary Detection of Randomly Occurring Signals,* Naval Electronics Laboratory Center, Technical Report TR-1741, November 1970 (30 pp.; unlimited distribution).
This report considers the application of a binary moving-window detector to the problem of detecting communication signals that appear at some unknown time and remain for a fixed time. Exact computational methods are given for obtaining detector performance statistics. A particular method of implementing the binary moving-window detector for detecting communication signals is given.

39. Dillard, R. A., "Detectability of Spread-Spectrum Signals," *IEEE Trans. on AES,* Vol. AES-15, No. 4, July 1979.
This paper describes methods of detecting spread-spectrum signals without knowledge of the PN code used to generate the signals. It gives exact and approximate methods for calculating detection performance for radiometers and for channelized receivers, the latter including application of the binary moving-window detector. Comparisons of pulse detection systems with systems that integrate energy over the transmission are shown to provide a basis for selecting signal parameters that minimize detectability.

Distribution-Free and Nonparametric Testing

40. Bell, C. B., *Automatic Distribution-Free Statistical Signal Detection,* Navy Electronics Laboratory Report 1245, 16 October 1964 (144 pp.; unlimited distribution).
Distribution-free statistical techniques are reviewed and evaluated for possible application to the signal detection problem.

41. Bradley, J. V., *Distribution-Free Statistical Tests,* Englewood Cliffs, NJ: Prentice-Hall, 1968 (388 pp.).
This well organized book gives detailed treatment of numerous distribution-free tests and examines their interrelationships. It is a good reference work, and the author's sense of humor makes it an entertaining textbook.

42. Carlyle, J. W., "Nonparametric Methods in Detection Theory," Chapter 8 in *Communication Theory,* edited by A. V. Balakrishnan, New York: McGraw-Hill, 1968, pp. 293–319.
The author considers the application of a number of existing nonparametric tests to several signal-detection situations, including those that involve colored noise.

43. Thomas, J. B., "Nonparametric Detection," *Proc. IEEE,* Vol. 58, No. 5, May 1970, pp. 623–631.
In this invited, tutorial paper, the author surveys the nonparametric approach to signal detection. He discusses and compares specific parametric and nonparametric detection schemes. The bibliography contains 37 references.

44. Dillard, G. M., *Distribution-Free Detection of Communication Signals,* Naval Electronics Laboratory Center Report TR 1776, September 1971 (24 pp.; unlimited distribution).
Two distribution-free procedures are described that provide a constant false-alarm rate despite changes in the noise environment. They are compared with optimum procedures that are based on the assumption that the probability distributions of the data are known.

45. Gibson, J. D., and J. L. Melsa, *Introduction to Nonparametric Detection with Applications,* New York: Academic Press, 1975 (241 pp.).
This book gives a tutorial development of nonparametric detection theory. It uses engineering terminology and stresses engineering applications. Only procedures suitable for applications to communication, radar, sonar, *et cetera,* are discussed. The differences between parametric and nonparametric detection are discussed, and examples of problems to which nonparametric detectors can be effectively applied are given.

46. Kassam, S. A., "A Bibliography on Nonparametric Detection," *IEEE Trans. on Inf. Theory,* Vol. IT-26, No. 5, September 1980, pp. 595–602.
A listing of papers from the engineering literature on nonparametric detection covering the period from 1959 to 1979 is given. The list is ordered alphabetically, and does not include reports of limited availability. In the brief introduction, the terms *nonparametric* and *distribution-free* are compared.

Compressive Receivers

47. Butson, P. C., *et al.,* "Theoretical Assessment of the Use of Pulse Compression in Panoramic Receivers," *Proc. IEE,* Vol. 113, May 1966, pp. 725–739.

This paper presents a theoretical investigation of the application of the pulse-compression technique to a panoramic superheterodyne receiver. It compares the sensitivity and resolution of such a system with those of a receiver without compression.

48. Harrington, J. B., and R. B. Nelson, "Compressive Intercept Receiver Uses SAW Devices for Signal Sorting with Stability and Flexibility," *Microwave Journal*, Vol. 27, No. 9, September 1974, pp. 57–62.

The design of a compressive receiver that uses three identical acoustic surface-wave dispersive filters is described. The receiver is designed to provide 100 KHz resolution over a 100 MHz bandwidth. The relationship between frequency resolution and sensitivity is discussed, and the probability of interception of pulses is evaluated. A simple, well illustrated description of the operation of a compressive receiver is given.

49. Maines, J. D., and E. G. S. Paige, "Surface-Acoustic-Wave Devices for Signal-Processing Applications," *Proc. IEEE*, Vol. 64, No. 5, May 1976; pp. 639–652.

Applications of surface-acoustic-wave (SAW) devices to signal processing are discussed. Basic information on SAW devices is given and devices relevant to signal processing are surveyed. Functional applications discussed include delay lines and memories, frequency filters, oscillators, and Fourier transformers. A section on spectral analysis includes a discussion of the implementation of filter banks, dispersive delay-line discriminators, and compressive receivers.

50. Hays, R. M., and C. S. Hartmann, "Surface-Acoustic-Wave Devices for Communications," *Proc. IEEE*, Vol. 64, No. 5, May 1976, pp. 652–671.

This paper reviews the current technology of SAW devices for application to communication systems. Applications discussed include bandpass filters, delay lines, and oscillators. It contains a section on spread-spectrum communication subsystems, and provides more than one hundred references on subjects related to SAW devices and their applications.

51. Milstein, L. B., and P. K. Das, "Spread Spectrum Receiver Using Surface Acoustic Wave Technology," *IEEE Trans. on Comm.*, Vol. COM-25, No. 8, August 1977, pp. 841–847.

This paper presents an overview of surface acoustic wave technology as applied to signal processing in spread-spectrum communication systems. Comparisons are made between SAW devices and other technologies with the conclusion that SAW technology offers the best alternative for high-bandwidth real-time operation.

52. Jack, M. A., *et al.*, "The Theory, Design, and Applications of Surface Acoustic Wave Fourier-Transform Processors," *Proc. IEEE*, Vol. 68, No. 4, April 1980, pp. 450–468.

This paper discusses the design and application of Fourier-transform processors that use SAW devices. The application of SAW devices to spectrum analysis, beam forming, and frequency-hopped waveforms is discussed. An extensive set of references is also given.

53. Nichols, D., "Transform Receivers for ECM Applications," *Microwave Journal,* Vol. 25, No. 10, October 1982, pp. 113–127.
An overview of the operating principles of compressive and acousto-optical receivers is given. Design trade-offs and development trends are discussed, with the conclusion that the compressive receiver is a reasonable alternative to instantaneous frequency measurement (1FM) receivers and channelized time-domain receivers.

High-Frequency (HF) Channels

54. Bello, P. A., "Error Probabilities due to Atmospheric Noise and Flat Fading in HF lonospheric Communications Systems," *IEEE Trans. on Comm.,* Vol. COM-13, September 1965, pp. 266–279. Reprinted in *Communications Channels: Characterization and Behavior,* edited by B. Goldberg, New York: IEEE Press, 1967.
Analytic expressions for the binary error rate of frequency-shift-keyed (FSK) and phase-shift-keyed (PSK) systems are derived. Several interesting facts are concluded about error probabilities under flat fading and various noise conditions and for different orders of diversity.
55. Goldberg, B., "300 kHz–30 MHz MF/HF," *IEEE Trans. on Comm.,* Vol. COM-14, December 1966, pp. 767–784. Reprinted in *Communications Channels: Characterization and Behavior,* edited by B. Goldberg, New York: IEEE Press, 1967.
The properties of the MF and HF bands are described in this tutorial paper. Error rate behavior is discussed for two adaptive communication systems.

Chapter 3
Calculation of Detection System Performance

This chapter contains formulas, instructions, and graphs for calculating the performance of energy-detection systems. The analysis is extensive only for the simpler detection systems, but instructions are given to some degree for all of the systems described in Chapter 2, except likelihood-ratio detection. When considering a particular signal structure, we suggest that the reader start with the section in Chapter 4 dealing specifically with that signal structure and then refer to the sections in this chapter that are referenced there. Regardless of the order used, we recommend a reading of Section 3.1 for an understanding of the detection model upon which calculations are based.

3.1 DEFINING THE DETECTION PROBLEM

Before computing the performance of an energy detector, we must specify the time-bandwidth product of the energy-integration cell and certain performance parameters. To do so, we need to decide what kind of system the interceptor is likely to use. Often, we must compare two kinds of systems (pulse-detection *versus* integration over the transmission) to decide which is most effective against our candidate signal. Conversely, we may be designing an intercept system for use against the communicator's signal. In this section, we discuss the steps involved in defining the energy-detection problem.

We assume that the interceptor knows everything about the signal except the pseudorandom "code-of-the-day" (see Section 1.3.). He knows what type of signal will be transmitted and the signal parameters listed in Table 1.1. However, most formulas in this chapter are valid even if the interceptor does not exactly match his system parameters to the signal parameters, provided all the signal energy is contained in the integration cell.

Before specifying a detection system, we must estimate to what extent the interceptor will dedicate his system to the signal in question. (For example, is he likely to match his system parameters to the signal parameters?) Also, we must estimate the detection-system *false-alarm rate* (FAR) that his sorting and recognizing procedures will tolerate.

A number of detection system types that use some form of energy detection were described in Chapter 2. The concept of dividing the time-frequency space under surveillance into time-bandwidth cells of energy integration was illustrated in Figure 2.13. Detectability calculations for pulse-detection systems use the cell concept, as described below, beginning at the system front end.

Step 1. For a pulse-detection system, specify the bandwidth W of the filters and the integration time T of the energy detectors. This defines a time-bandwidth grid of integration cells, each of size $T \times W$. The cell size is usually chosen to match the size of data symbols (more correctly, data-symbol signals), elements of data symbols, or other distinct signals composing the transmission. We assume an ideal "fit" of the grid to the signal, resulting in only two kinds of integration cells, those containing signal energy (each the same amount) and empty ones (empty except for noise). In the case of energy integration over the entire transmission, there is just one cell with signal.

Step 2. For integration cells occupied by signal energy, estimate the SNR at the input to the interceptor's receiver. Adjust this ratio appropriately to account for radiometer losses, nonwhite noise, splitting of the signal in time or frequency (i.e., into more than one time-bandwidth cell), or any other factor that differs from the ideal detection model assumed. Some instructions for doing so are given in Section 3.3.

Step 3. For a specified value of false-alarm probability per cell and for the adjusted SNR (from Step 2) for a cell with signal, calculate or estimate from graphs the performance of an energy detector under the assumption of ideal energy detection in stationary, white Gaussian noise.

Step 4. Consider the detection system as a whole, and evaluate the performance (detection probability and FAR) of the *alarm circuit* as a function of the inputs to it from the energy detectors, one input from each time-bandwidth cell. (See Figure 2.2 and the discussions at the beginning of Chapter 2 on alarm-circuit decision procedures.) Later sections in this chapter give calculation instructions for various types of alarm-circuit decision procedures.

More often, we perform Steps 2–4 in reverse order. We specify system performance and work backward to estimate the signal power required per integration cell with signal. Although exact calculations are difficult, good estimates usually can be found easily by using the various graphs and approximations given in this chapter.

3.2 ENERGY-DETECTION CALCULATIONS

For the analysis of energy detection, we assume that the noise at the input to the energy detector is a zero-mean, stationary, Gaussian random process that has a flat, band-limited power spectral density. In particular, the model assumed here (the F-ED-T system in Figure 2.9) includes an ideal rectangular filter of bandwidth W preceding the energy detector. The normalized decision statistic V' is related to the integrated energy V by $V' = 2V/N_{01}$, where N_{01} denotes the one-sided noise power spectral density. The thresholder compares the value of V' with a threshold K, and the decision is "signal present" if $V' \geq K$, and "noise only" if $V' < K$.

When TW is in the region of unity, the received signal energy depends on the pulse shape and filter shape. For the detection of rectangular and Gaussian-shaped pulses, and for a number of filter shapes, Skolnik [1, p. 374] tabulates the optimum time-bandwidth product (ranging from 0.4 to 1.7) and the loss in SNR compared with a matched filter. Also, Schwartz [2, p. 419] gives similar information in graphical form for rectangular pulses. For larger TW, we assume an ideal rectangular filter; the shape of the signal does not matter if it is contained in the time-bandwidth area of integration. This is discussed further when properties of the noncentral chi-square distribution are described.

The operation of the thresholder depends on the type of energy detector assumed. The terms *continuous integration* and *integrate-and-dump* (I+D) were defined in Section 2.1.1. For continuous integration, V' is the value of the normalized integrated energy:

$$V'(t) = \frac{2}{N_{01}} \int_{t-T}^{t} y^2(r)\, \mathrm{d}r \tag{3.1}$$

at any particular instant t. For I+D detection, the decision times are at instants T seconds apart, and V' represents the value of $V'(t)$ only at one of these particular instants. In either case, if the value of the test statistic V' exceeds the threshold K, the decision (for that particular energy detector) at that instant is "signal present." The false-alarm probability per decision is

$$Q_F = \text{Prob}[V' \geq K \text{ at time } t \mid \text{no signal energy integrated during } (t - T, t)],$$

and the detection probability per decision is

$$Q_D = \text{Prob}[V' \geq K \text{ at time } t \mid \text{signal energy } E_S \text{ integrated during } (t - T, t)].$$

The vertical bar indicates that the probability is conditional on the event that follows it.

For now, we consider the time-bandwidth space to be divided into cells of size

$T \times W$ and disregard the time-alignment problem (which is the alignment of the signal with the integration interval). We treat Q_F as the probability of a false alarm in an empty time-bandwidth cell and Q_D as the detection probability for a cell containing signal. Adjustments to detectability estimates for continuous integration or I+D are described in Section 3.3.

3.2.1 Exact Calculations

The distribution function of the test statistic V' is closely approximated by the chi-square distribution with $2TW$ degrees of freedom in the noise case and by the noncentral chi-square distribution with $2TW$ degrees of freedom and noncentrality parameter $\lambda = 2E_S/N_{01}$ in the signal case [3; 4, p. 228]. Therefore, Q_F and Q_D are given by

$$Q_F = \int_K^\infty \frac{x^{TW-1} e^{-x/2}}{2^{TW}\Gamma(TW)} \mathrm{d}x \tag{3.2}$$

and

$$Q_D = \int_K^\infty \frac{1}{2}\left(\frac{x}{\lambda}\right)^{(TW-1)/2} e^{-(x+\lambda)/2} I_{TW-1}(\sqrt{x\lambda})\mathrm{d}x \tag{3.3}$$

In (3.3) the function $I_\gamma(z)$ is the modified Bessel function of the first kind, order γ. Because the noncentral chi-square distribution closely approximates the true distribution of the energy-detector output (see appendix of [3]), results obtained by using (3.2) and (3.3) will be called *exact* results.

The noncentral chi-square distribution with n degrees of freedom is the distribution of the sum of squares of n independent normal random variables, each with unit variance and at least one with nonzero mean. (If all have zero mean, the distribution is chi-square.) The noncentrality parameter λ is the sum of squares of the means. Therefore, there is no requirement that the means are equal; the same distribution is obtained for any sequence of means with squares that sum to λ. This means that the total energy is important, but how it is distributed over the time-bandwidth cell of energy integration is not.

By making a change of variables $u = x/2$, $S = \lambda/2$, and $K_0 = K/2$, (3.2) and (3.3) become

$$Q_F = \int_{K_0}^\infty \frac{u^{TW-1} e^{-u}}{\Gamma(TW)} \mathrm{d}u \tag{3.4}$$

and

$$Q_D = \int_{K_0}^{\infty} \left(\frac{u}{S}\right)^{(TW-1)/2} e^{-(u+S)} I_{TW-1}(2\sqrt{uS})\mathrm{d}u \tag{3.5}$$

For $TW = 1$, the integrand in (3.5) reduces to the square-law form of the familiar Rice (or Rician) distribution that describes envelope detection of a single pulse. In the more general case (with TW an integer), the equation for Q_D in (3.5) is the generalized Q-function, $Q_D = Q_{TW}(\sqrt{2S},\sqrt{2K_0})$, where $S = E_S/N_{01}$ is the SNR [5]. Equation (3.5) is also known as the Mth-order Q-function, with $M = TW$ [6]. In the following we assume that TW is an integer. We note for later use that the mean and variance of the random variable u defined just prior to (3.4) are [7, p. 943]:

$$\begin{aligned} \mu &= TW + S \\ \sigma^2 &= TW + 2S \end{aligned} \tag{3.6}$$

The value of the normalized threshold, or bias level, K_0 that provides the desired Q_F is found in Pachares' table [8] for $TW = 1, 2, \ldots, 150$, and for $Q_F = 10^{-P}$, $P = 1, 2, \ldots, 12$. A series expansion of (3.4), obtained by integration by parts, can be used to compute K_0 directly. Alternatively, the techniques outlined by Helstrom [6] provide an approximation to K_0; the approximation usually suffices for moderately large values of TW (e.g., for $TW > 20$). Torrieri (Equation (4.2.42) of [4]) provides a simple approximation relating the false-alarm probability and the threshold.

Evaluation of (3.5) to determine Q_D can be accomplished in any of several ways. Robertson [9] and Dillard [5] give recursive formulas that can be used when the parameters result in numerical stability. Parl [10] also provides a recursive formula that is numerically stable and simple to program. Helstrom [6] gives techniques for an approximate evaluation of (3.5). The curves presented later were computed by using Robertson's formula for $S < 10^{2.3}$ (23 dB SNR); otherwise, Helstrom's technique [6] or an approximation based on the normal distribution was used. This approximation and others are discussed in the next section.

Most radar texts and handbooks have graphs that can be used to determine energy detection performance, but not always for the desired parameters. These graphs usually are for the noncoherent integration of n independent, square-law detected, radar pulses, each with the same SNR ρ. To use these graphs, set $TW = n$ and $\rho = S/n$. An exception to this notation is found in DiFranco and Rubin [11]. Their graphs are in terms of the peak SNR, rather than the average SNR ρ. Therefore, to use their results, let the peak SNR be $S/2n$. A word of caution about this difference is given by DiFranco and Rubin [11, pp. 296–298].

3.2.2 Energy Detection Performance Approximations

We first illustrate a method of estimating energy detection performance by calculating the SNR required to detect a pulse having a unity time-bandwidth product and adjusting the result with an estimate of integration loss. The term *integration loss* is used in radar literature to quantify the difference between integrating radar pulses noncoherently and integrating them coherently. Recall from Section 3.2.1 that using noncoherent integration to detect n radar pulses is mathematically equivalent to using energy detection over time-bandwidth area $TW = n$ to detect the same total amount of signal energy. Coherently integrating n radar pulses by using phase information is ideally equivalent to detecting a single pulse that has n times greater energy.

Figure 3.1 (a) gives the single-pulse detection probability (Q_D for $n = TW = 1$) as a function of the SNR for various values of the false-alarm probability Q_F. The SNR ρ is the value of S for $TW = 1$. Figure 3.1 (b) is similar to one of Barton's [12, Figure 7], and gives the integration loss for the noncoherent integration of n pulses compared with coherent integration of n pulses. These curves were computed by averaging the required E_S/N_{01} over a range of combinations of Q_F and Q_D for the specified values of ρ and TW. Figure 3.1 gives instructions for using the curves in (a) and (b) to find estimates of the required S for specified energy-detection performance.

Basically, we find the single-pulse SNR ρ for a given Q_F and Q_D. This value of ρ is n times greater than the per-pulse SNR when n pulses are coherently integrated, so we add to it (in dB) the integration loss, corresponding to TW and ρ, to find the value of S required for noncoherent integration. Barton [12] calculated the integration loss by using an empirically derived relationship between detector loss and the IF SNR. Cann [13] further simplified Barton's procedure for estimating detector loss. He presented a single curve rather than Barton's family of curves, at the expense of slightly poorer accuracy.

Urkowitz [14] modified Barton's detector-loss relationship to find a closed-form expression for noncoherent integration gain. By adapting Urkowitz's expression to the problem of estimating the S required for a specified TW, we have the formula:

$$S \approx \frac{4.6TW}{-1 + (1 + 9.2TW(\rho + 2.3)/\rho^2)^{1/2}} \tag{3.7}$$

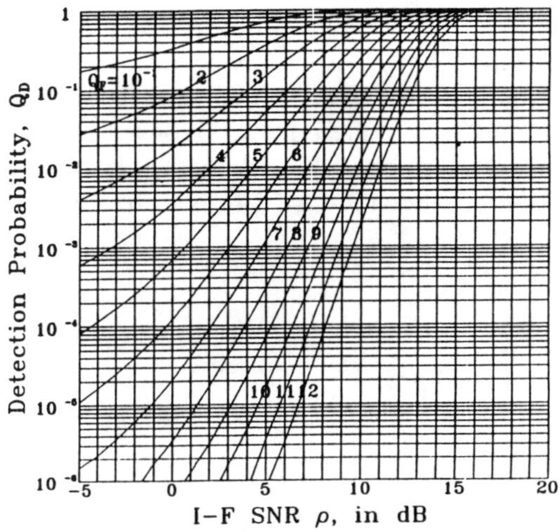

Figure 3.1 Graphs for finding E_s/N_{01}, given TW, Q_D, and Q_F: (a) single-pulse case; (b) noncoherent integration loss (after Barton [12]). Find in (a) the value of ρ corresponding to Q_F and Q_D; find in (b) the integration loss L_i; let $E_S/N_{01} = \rho + L_i$. If $TW > 10^4$, let $L_i = (L_i \text{ for } TW = 10^4) + 5 \log_{10}TW - 20\text{dB}$.

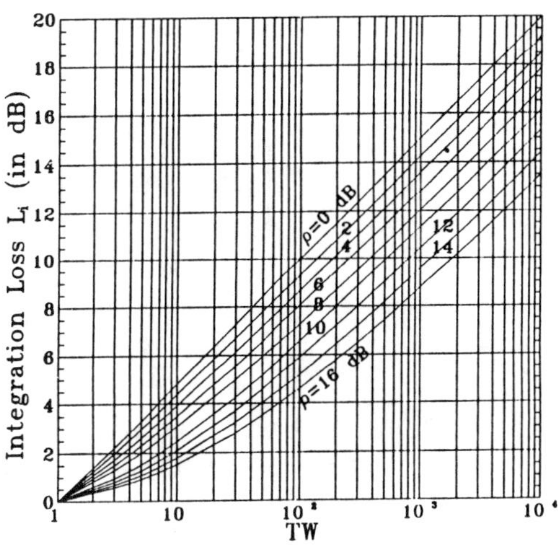

Figure 3.1b

where ρ is the IF SNR in Figure 3.1 (a) (not in dB) for the single-pulse case (i.e., for $TW \approx 1$). (Recall that $S \equiv E_S/N_{01} = n\rho$, where E_S is the signal energy in the time-bandwidth cell having area TW.) Some sample calculations of this approximation through $TW = 10^8$ show the error to be less than ± 0.4 dB [15]. By using another closed-form expression given by Urkowitz [14], we are able to find Q_D given S and TW. By rewriting his Equation 4 (with his Equation 2) in terms of

$$U = (S + 2.3TW)/S^2 \tag{3.8}$$

we have [15]:

$$\rho \approx \left(1 + \sqrt{1 + 9.2U}\right)/2U. \tag{3.9}$$

This estimate of ρ is used in Figure 3.1 (a) to find Q_D.

For large values of TW, the central limit theorem [11, p. 96] can be used to obtain normal approximations of Q_F and Q_D. (See Section V of Urkowitz [3].) In terms of the normal *cumulative distribution function* (CDF):

$$F(y) = \frac{1}{\sqrt{2\pi}} \int_{-\infty}^{y} e^{-z^2/2} \mathrm{d}z \tag{3.10}$$

the approximations are

$$Q_F \approx 1 - F\left[(K_0 - TW)/\sqrt{TW}\right] \tag{3.11}$$

and

$$Q_D \approx 1 - F\left[(K_0 - TW - S)/\sqrt{TW + 2S}\right] \tag{3.12}$$

Equations (3.11) and (3.12) are obtained by using (3.6) and applying the central limit theorem. By solving (3.11) for the normalized threshold K_0 we obtain

$$K_0 \approx F^{-1}(1 - Q_F)\sqrt{TW} + TW, \tag{3.13}$$

where $F^{-1}(x)$ is equal to the variate y such that $F(y) = x$. Substitution of (3.13) into (3.12) yields

$$Q_D \approx 1 - F\left\{[F^{-1}(1 - Q_F) - S/\sqrt{TW}]/\sqrt{1 + 2S/TW}\right\}. \tag{3.14}$$

The approximation in (3.14) is referred to as the *full normal approximation*. Another approximation is made by noting that for $S \ll 2TW$,

$$Q_D \approx 1 - F[F^{-1}(1 - Q_F) - S/\sqrt{TW}]. \qquad (3.15)$$

We can safely assume a normal distribution for some of the detection systems described in Chapter 2 only when certain constraints are met; these constraints vary among systems. The approximations given by (3.14) and (3.15) for energy detection require the product TW to be large, but their accuracy also depends on the values of Q_F and Q_D. Equation (3.12) is a good approximation to Q_D when K_0 is the *true* bias level in (3.4), even for relatively small values of TW (TW larger than about 20). However, a comparison of the approximations to K_0 in (3.13) with the true values in Pachares' table [8] shows that (3.14) is in error primarily because the approximation to K_0 (3.13) is in error. The same situation exists for TW outside the range in Pachares' table; that is, for $TW > 150$.

If K_0 is available, we also can use (3.12) to estimate the SNR required to achieve a given value of Q_D. The estimate is

$$\hat{S} = K_0 - TW + \gamma^2 - \gamma\sqrt{2K_0 - TW + \gamma^2} \qquad (3.16)$$

where

$$\gamma = F^{-1}(1 - Q_D). \qquad (3.17)$$

When $Q_D = 0.5$, $\gamma = 0$, and the estimate is

$$\hat{S}(0.5) = K_0 - TW \qquad (3.18)$$

Equation (3.16) is obtained by solving the quadratic equation $K_0 - TW - S = \gamma\sqrt{TW + 2S}$ for S.

Figure 3.2 provides a means for quickly estimating K_0 when Q_F and TW are given ($TW \le 2000$). Figure 3.2 (a) essentially is a graph of Pachares' table [8] for odd values of the power of ten that defines Q_F. Figure 3.2 (b) gives the difference $K_0 - TW$ as a function of TW for $Q_F = 10^{-k}$, $k = 3, \ldots, 12$. By using (3.18), we see that the curves in Figure 3.2 (b) provide estimates of the required SNR whenever $Q_D = 0.5$.

The two normal approximations (3.14) and (3.15) and the "hybrid" approximation (3.12) using the true K_0 are compared with the "exact" results (3.5) in Figure 3.3 for $Q_F = 10^{-12}$ and in Figure 3.4 for $Q_F = 10^{-4}$. By inspecting these two figures, we see that (3.12) gives results essentially equal to (3.5), except for small TW (when none of the approximations is reasonable). Thus, we conclude that if the true values of the normalized threshold K_0 are available (e.g., either tabulated [8], or computed

[6]), there is no reason for using (3.14) or (3.15). In fact, Hasting derives approximations [7, p. 932] that make evaluating (3.12) much simpler than evaluating (3.5) by using Robertson's [9] or Helstrom's [6] techniques.

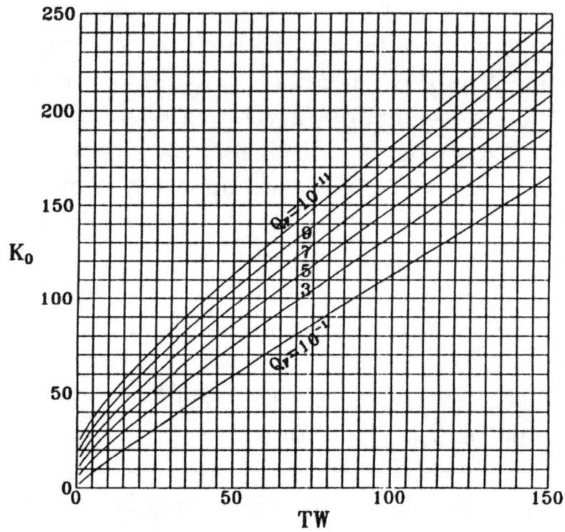

Figure 3.2 Quantization threshold K_0 for values of the false-alarm probability: (a) curves matching exact values in [8]; (b) threshold less TW for a larger range of TW. For $Q_D = 0.5$ and $TW > 200$, the required E_S/N_{01} is approximately $K_0 - TW$.

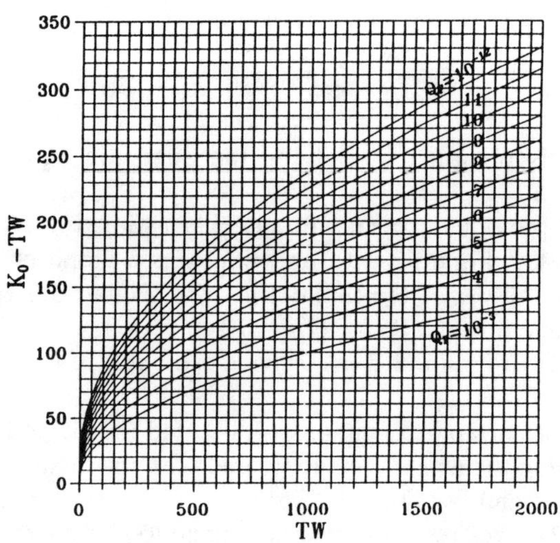

Figure 3.2b

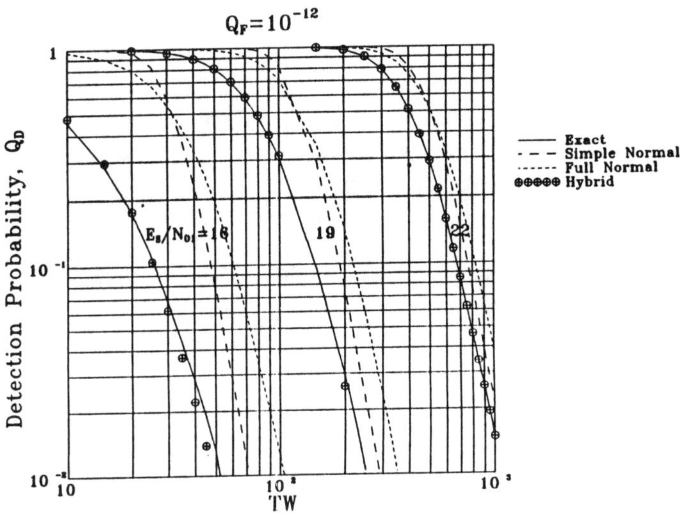

Figure 3.3 Comparison of approximations with exact distribution ($Q_F = 10^{-12}$).

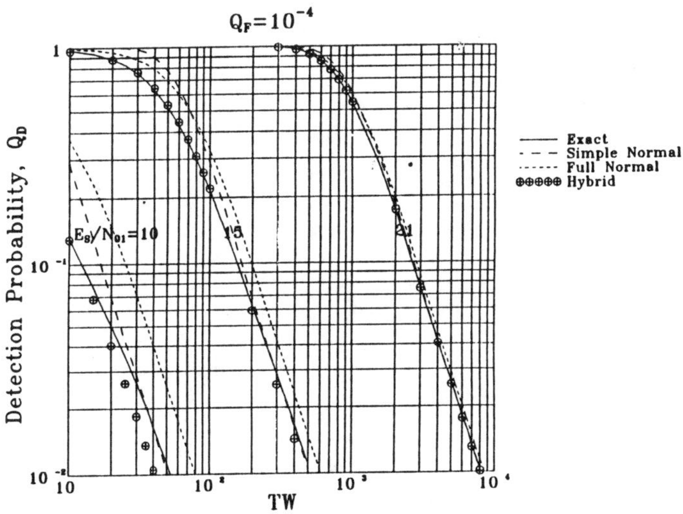

Figure 3.4 Comparison of approximations with exact distribution ($Q_F = 10^{-4}$).

3.2.3 Graph Set I

Graph Set I (Figures 3.5 to 3.13) gives curves of Q_D *versus* TW for $Q_F = 10^{-12}$, 10^{-11}, ..., 10^{-4}. As indicated above, Robertson's algorithm [9] was used to compute Q_D for $S < 10^{2.3}$ (23 dB); otherwise, Helstrom's technique [6] or the "hybrid" approximation described above was used.

When using Graph Set I for the time-collapsing system F-ED-C/T, let the system integration time be equal to the product of the energy-detector integration time T and the number of taps. Further instructions are given in Section 3.4.

GRAPH SET I
(Radiometer Curves)

Detection probability Q_D per cell containing signal *versus* time-bandwidth product TW of cells for constant per-cell false-alarm probability Q_F with curves given for values of E_s/N_{01}:

E_s = signal energy in a cell containing signal;
N_{01} = one-sided noise power spectral density.

Figure	Q_F
3.5	10^{-12}
3.6	10^{-11}
3.7	10^{-10}
3.8	10^{-9}
3.9	10^{-8}
3.10	10^{-7}
3.11	10^{-6}
3.12	10^{-5}
3.13	10^{-4}

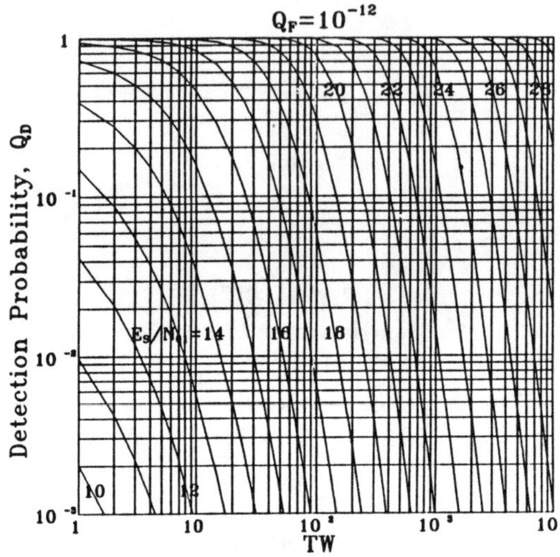

Figure 3.5 Per-cell detection probability for $Q_F = 10^{-12}$.

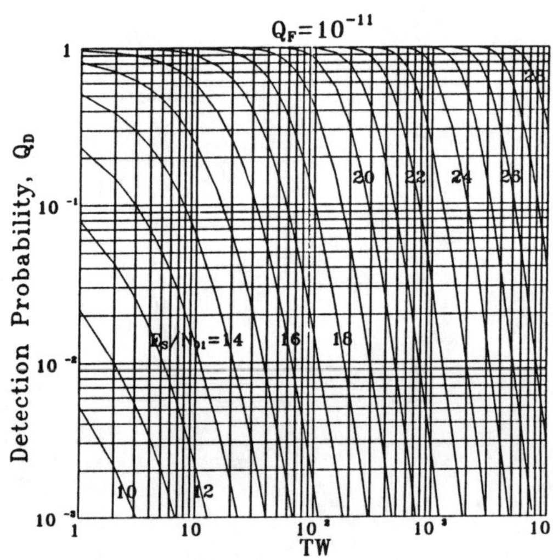

Figure 3.6 Per-cell detection probability for $Q_F = 10^{-11}$

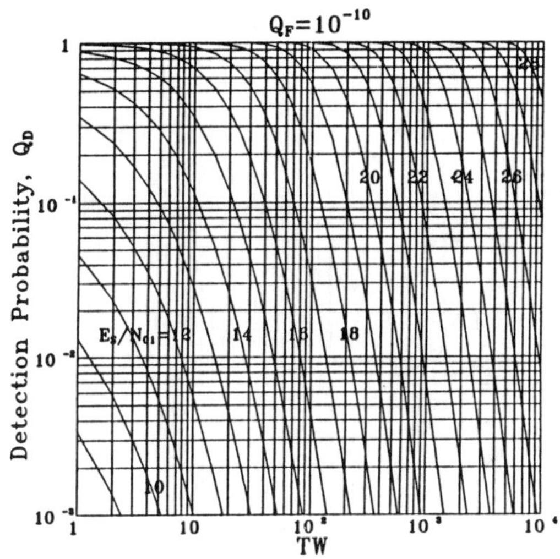

Figure 3.7 Per-cell detection probability for $Q_F = 10^{-10}$

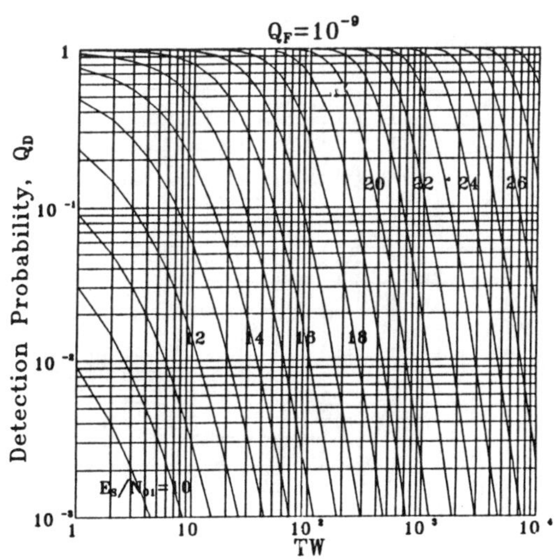

Figure 3.8 Per-cell detection probability for $Q_F = 10^{-9}$.

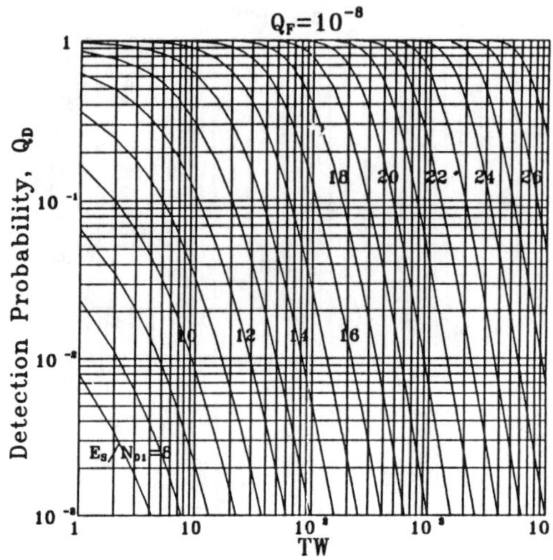

Figure 3.9 Per-cell detection probability for $Q_F = 10^{-8}$.

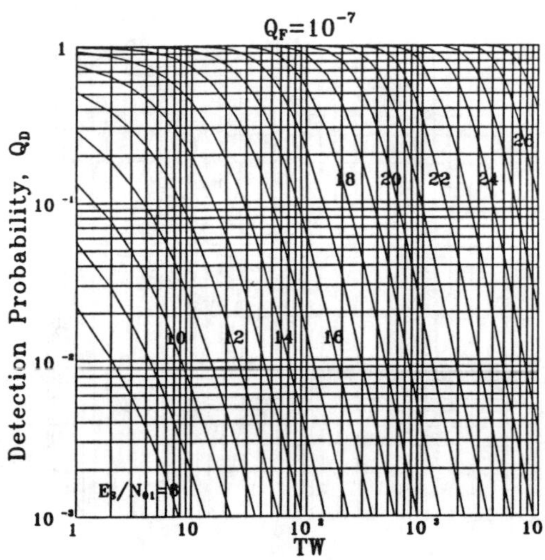

Figure 3.10 Per-cell detection probability for $Q_F = 10^{-7}$.

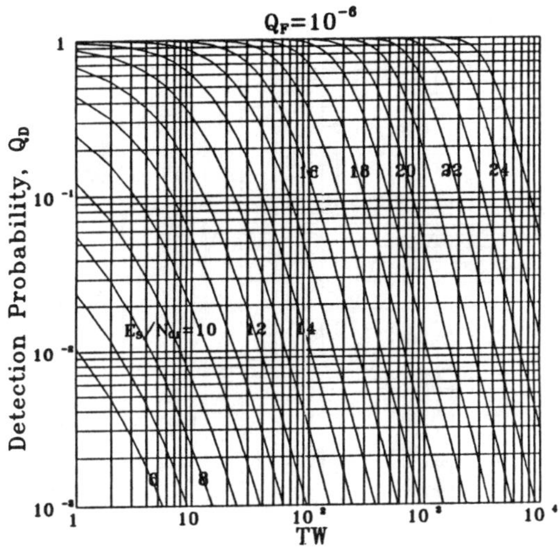

Figure 3.11 Per-cell detection probability for $Q_F = 10^{-6}$.

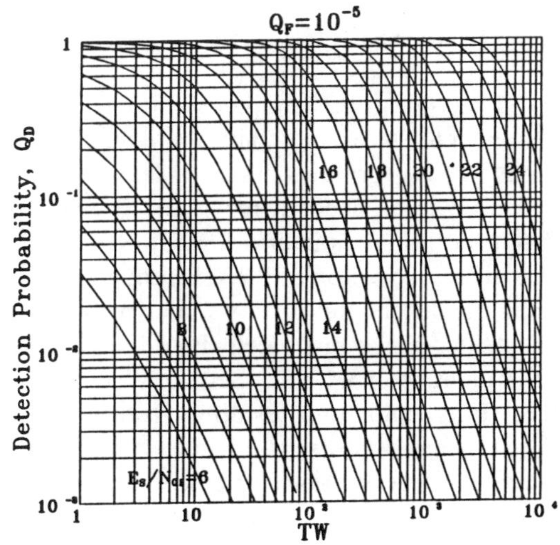

Figure 3.12 Per-cell detection probability for $Q_F = 10^{-5}$.

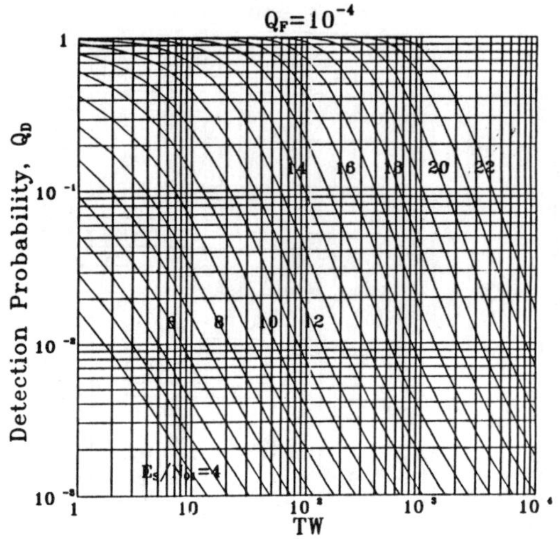

Figure 3.13 Per-cell detection probability for $Q_F = 10^{-4}$.

3.3 ADJUSTMENTS FOR LOSSES

The formulas for signal detection in this chapter are confined to the detection problem beginning at the interceptor's receiver with the exception of some discussions on antenna considerations. These formulas assume ideal energy detection in stationary, white Gaussian noise. We can account for some receiver losses by subtracting a constant (dB loss) from the initial estimate of SNR at the receiver. For example, radiometer constants are given in Section 3.3.2. The equations and graphs for energy detection assume rectangular filters for $TW \gg 1$ and matched filters for $TW = 1$; any losses resulting from filter-shape mismatch should also be taken into account in the initial estimate of the interceptor's SNR. (See the discussion in Section 3.2 concerning filter mismatch.) Interceptor loss resulting from not knowing precisely the possible frequencies of FH signals is discussed in Section 3.3.3, as is I+D loss.

We should consider the effect of the environment not only on the signal strength, but also on the statistics of signal and noise at the input to the interceptor's receiver. Major losses occur at frequencies where our assumption of stationary, white, Gaussian noise does not apply, and we discuss this next.

3.3.1 Losses at Lower Frequencies

In Section 2.5.1, we discussed signal fading and external noise, and described some methods of reducing their effects. In a signal-fading environment, the equations given later in Sections 3.4 and 3.5 for the detection probability P_d of pulse-detection systems are not accurate because they assume identical statistics for each pulse and pulse-to-pulse statistical independence.

We know of no simple way to estimate detectability accurately in colored or nonstationary noise. Distribution-free and nonparametric detection techniques are appropriate in these cases (see Section 2.5). To estimate detectability crudely, we can assume that the effect of external noise and fading is the same as any loss that causes a fixed decrease (in dB) in the interceptor's SNR, with the size of the loss depending on the type and amount of noise and interference.

3.3.2 Radiometer Constants

Some basic types of radiometers were described and illustrated in Section 2.1.2. The constants associated with these radiometers (from Tiuri [16]) are listed in Table 3.1, with L_r representing the constants in dB. The annotated bibliography on radiometers in Chapter 2 will guide the interested reader to discussions on a number of other radiometer types. In detectability calculations the effective energy is found by subtracting the loss L_r (and other losses) from the estimate of signal energy (in dB) at the interceptor's antenna output terminal. If the radiometer uses digital integration, the associated degradation should be included in the loss L_r. Curves of the degradation factor for digital integration are given by Ohlson [17].

Table 3.1 Radiometer Constants

Radiometer	*Figure*	*Degradation Factor*
Total Power	2.6	1 (L_r = 0 dB)
Dicke, square-square	2.7	2 (L_r = 3.01 dB)
Dicke, square-sine	2.7	$\pi/\sqrt{2}$ (L_r = 3.47 dB)
Dicke, sine-sine	2.7	$2\sqrt{2}$ (L_r = 4.52 dB)
Correlation receiver (two receivers and one antenna)	2.8	$\sqrt{2}$ (L_r = 1.51 dB) (When receiver noise is much greater than the antenna background noise.)

3.3.3 Signal-Splitting Losses

In Section 3.2, we defined Q_D as the detection probability per time-bandwidth cell of energy integration. To evaluate the effects of signal-splitting losses, we express Q_D as a function of E_S and define $Q_D(E_S)$ as the probability of a signal decision in a time-bandwidth cell (size TW) containing signal energy E_S. Also, we assume that the frequency band of integration covers the signal band, the integration time is approximately equal to the signal duration, and the signal energy is approximately evenly distributed over its duration. Also, the signal might consist of time-hopped pulses, for example, or it might be a spread-spectrum pulse. When continuous integration (Figure 2.4) is used, at some time, an integration interval is aligned with the signal, and $Q_D(E_S)$ is slightly less than but close to the true probability of detecting the signal. When nonoverlapping ID detection is used, lack of knowledge about arrival time results in the time-alignment problem illustrated by Figure 3.14.

When not aligned in time, the signal arrives at the I+D detector at some random time during an integration interval and ends during the next one; thus, the signal is split between two time-bandwidth cells. The probability of detecting the signal is the probability that a signal decision occurs in at least one of the two time-bandwidth cells containing signal, averaged over all possible starting times. This detection probability is given by

$$\bar{Q}_D = \bar{Q}_D(E) = 1 - 2\int_{.5}^{1} [1 - Q_D(xE)][1 - Q_D(E - xE)]\mathrm{d}x \qquad (3.19)$$

where E is the interceptor's received energy, usually the energy E_1 of the entire transmission or E_2, the energy of a data symbol.

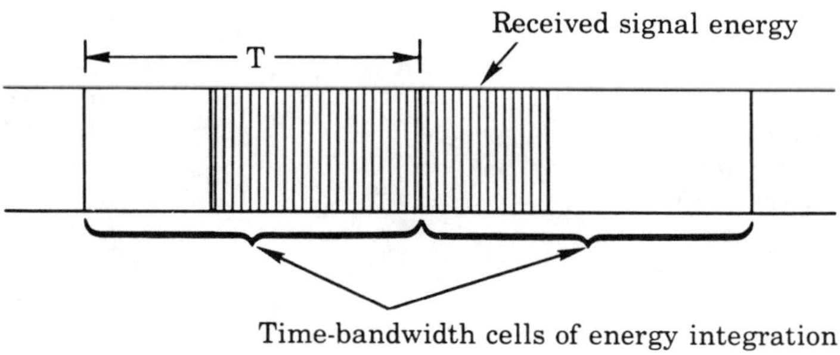

Figure 3.14 Integrate-and-dump (I + D) detection.

The probabilities involved in the integral in (3.19) are illustrated in Figures 3.15 and 3.16 for $TW = 1$ and $TW \gg 1$. Note that the probability of a signal decision in the cell containing less signal has a significant effect on the probability of detecting the signal only when TW is small. Generally, the probability is dominated by the contribution from the cell with more signal, and the approximation

$$\bar{Q}_D(E) \approx 2\int_{.5}^{1} Q_D(xE)\mathrm{d}x \tag{3.20}$$

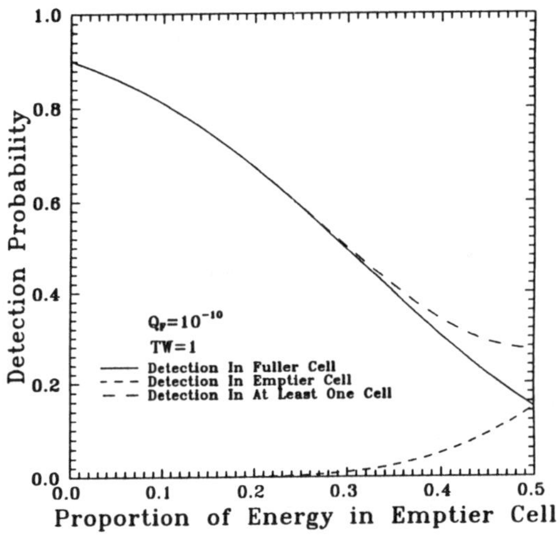

Figure 3.15 Example of splitting signal energy into two time intervals of energy integration ($TW = 1$).

is valid when $TW \gg 1$ or $Q_D(E) < 0.5$.

A fairly accurate detection estimate, which also avoids the calculation of $\bar{Q}_D(E)$ (or of E given $\bar{Q}_D$), can be made by using the effective loss L_t. This estimate is

$$\bar{Q}_D(E) = Q_D(E - L_t) \tag{3.21}$$

with E and L_t in dB.

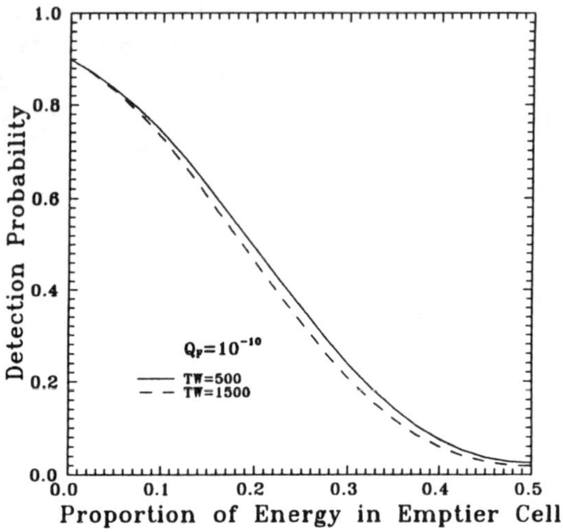

Figure 3.16 Example of splitting signal energy into two time intervals of energy integration ($TW \gg 1$).

Figure 3.17 gives curves of the signal-splitting loss L_t *versus* detection probability for $TW = 1$ and $TW = 10^4$, and for $Q_F = 10^{-10}$ and $Q_F = 10^{-6}$. The detection probability is the average over the distribution of the proportion of energy in the emptier cell. The data for these curves was obtained as follows. For each combination of TW and Q_F, an initial value of E_S/N_{01} was chosen. The proportion of energy in the emptier cell was incremented from 0.0 to 0.5, the probability of detection in at least one cell was computed, and the results were (numerically) averaged to obtain $\bar{Q}_D$. The (smaller) E_S/N_{01} required to achieve $Q_D = \bar{Q}_D$ when there is no splitting was then computed by using (3.5). The difference (in dB) between this smaller E_S/N_{01} and the one used above is the splitting loss shown in Figure 3.17. The process was repeated, by incrementally increasing the (split) E_S/N_{01}, until values of $\bar{Q}_D$ over the range shown were obtained. To estimate the received signal energy E required by an I+D energy detector for a given "true" detection probability $\bar{Q}_D$, find E_S/N_{01} (in dB) for $Q_D(E_S) = \bar{Q}_D$ (in Graph Set I or later graphs) and let $E/N_{01} = E_S/N_{01} + L_t$.

We next consider the loss due to splitting a signal in frequency, a problem of interest primarily when estimating the vulnerability of FH signals to a filter-bank system with nonoverlapping, contiguous passbands. For now, we assume that the signal energy (of the data symbol or pulse) is contained entirely within a time-interval of energy integration and only consider the possibility of its being split into two filter passbands. To make the analysis tractable, we assume that the filters have rectangular passbands with bandwidths equal to the signal bandwidth. We further assume that the energy of each signal is approximately evenly distributed over its frequency band.

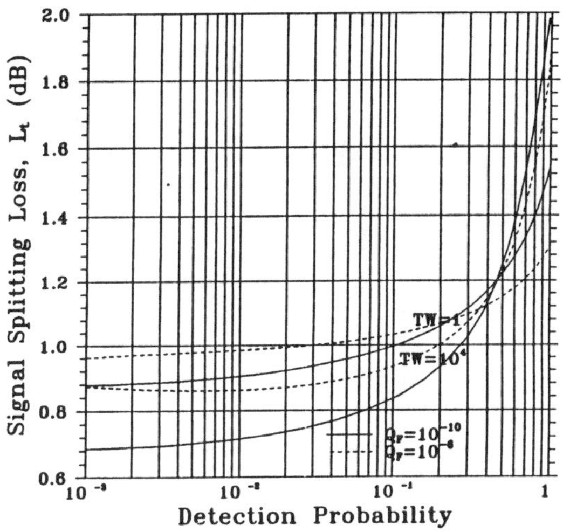

Figure 3.17 The effective energy loss resulting from a random splitting of the signal energy in time into two cells of integration.

(The analyses for I+D detection and for filter-bank detection cannot both be accurate for small values of TW, because an even distribution of energy in both time and frequency is then impossible.) Under these conditions, the problem of splitting in frequency is exactly analogous to the problem of splitting in time. The true detection probability $\bar{Q}_D(E)$ is given by (3.19), and Figure 3.17 is also valid (with the time-bandwidth cells contiguous in frequency rather than in time). By defining the effective loss L_f due to splitting in frequency:

$$\bar{Q}_D(E) = Q_D(E - L_f) \tag{3.22}$$

(with E and L_f in dB), we can use Figure 3.17 simply by replacing L_t with L_f. To estimate the received energy E required (before splitting in frequency) for a given true detection probability $\bar{Q}_D$, find E_S/N_{01} (in dB) for $Q_D(E_S) = \bar{Q}_D$ in Graph Set I or in later graphs, and let $E/N_{01} = E_S/N_{01} + L_f$.

The loss L_t is smaller than that given in Figure 3.17 if the signal duration is less than the integration time and, analogously, the loss L_f is smaller if the signal bandwidth is less than the filter bandwidth. However, in both cases the amount of noise integrated would then be unnecessarily large, reducing the detection probability.

The assumptions made in the analyses of splitting in time and in frequency are somewhat restrictive, especially in the frequency-splitting case. The actual loss L_f will vary with the true spectral shapes of the filter and signal.

Finally, we consider the splitting loss that occurs with the combination of I+D detection and overlapping filters. When the signal can be split only in time or only in frequency, the uniform distribution describes the amount of signal energy in the fuller time-bandwidth cell, under the assumptions stated earlier. When the signal can be split both in time and in frequency, the distribution of the proportion y of energy in the fullest cell (of the four possible cells) is

$$F(y) = \begin{cases} 0, & y \leq \frac{1}{4} \\ 4y(\ln[4y] - 1) + 1, & \frac{1}{4} < y \leq \frac{1}{2} \\ 4y(1 - \ln[y]) - 3, & \frac{1}{2} < y \leq 1 \\ 1, & 1 < y \end{cases} \tag{3.23}$$

Equation (3.23) assumes an even distribution of signal energy in time and frequency, rectangular passbands, and both T and W matched to the signal dimensions.

By differentiating (3.23) with respect to y, we find the density of y:

$$f(y) = \begin{cases} 0, & y \leq \frac{1}{4} \quad \text{or} \quad y > 1 \\ 4 \ln 4y, & \frac{1}{4} < y \leq \frac{1}{2} \\ -4 \ln y, & \frac{1}{2} < y \leq 1 \end{cases} \tag{3.24}$$

The average detection probability $\bar{Q}_D(E)$ is obtained from

$$\bar{Q}_D(E) \approx \int_{1/4}^{1} Q_D(E|y) f(y) dy \tag{3.25}$$

where $Q_D(E|y)$ is the probability of detection in at least one of the four possible cells containing the signal, conditional on having the proportion of energy y in the fullest cell. By specifying E_S/N_{01} and using a procedure similar to the one described for computing the curves in Figure 3.17, we obtained estimates of $Q_D(E|y)$ for $1/4 < y < 1$. We used these estimates and (3.24) in (3.25), to perform the numerical integration. We then found the (smaller) required E_S/N_{01} such that $Q_D = \bar{Q}_D(E)$ when there is no splitting. The difference between this smaller value and the one used above is the splitting loss L_{t+f}. We repeated this procedure for several choices of E_S/N_{01} to obtain a range of values of Q_D.

Figure 3.18 shows L_{t+f} and L_t (or L_f) *versus* the detection probability for $TW = 10^4$ and $Q_F = 10^{-6}$ and $Q_F = 10^{-10}$. The solid squares in this figure indicate that $L_{t+f} \approx 2L_t$ (or $L_{t+f} \approx L_t + L_f$). These results for $TW = 10^4$ show that the loss in dB is approximately twice as great as L_f or L_t for any particular value of $\bar{Q}_D(E)$. For use in vulnerability calculations, we generalize these results to

$$Q_D(E) = Q_D(E - L_t - L_f) \tag{3.26}$$

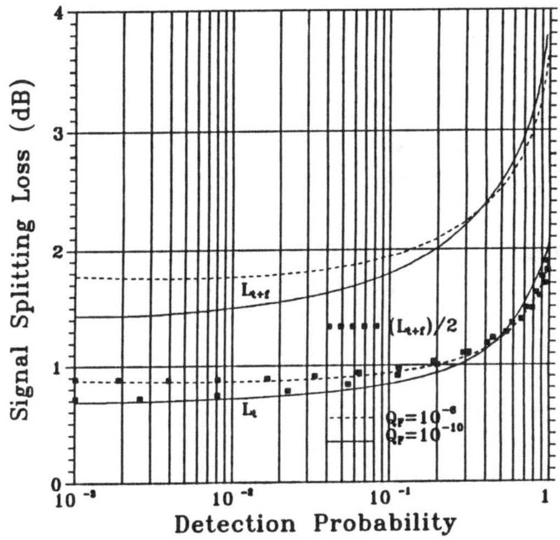

Figure 3.18 The effective energy loss L_{t+f} resulting from a random splitting of the signal energy in time and frequency into four cells of integration, for $TW = 10^4$. Points for $L_{t+f}/2$ and for curves of L_t (or L_f) indicate that splitting losses for frequency and time are additive.

which is the same as (3.21) when $L_f = 0$, and the same as (3.22) when $L_t = 0$. Given TW and a specified detection probability $\bar{Q}_D$, find $L_t = L_f$ in Figure 3.17 (if the assumptions are approximately satisfied); use one of the graphs in Graph Set I (or in Section 3.4) to find the value of E_S/N_{01} (in dB) corresponding to $Q_D(E_S) = \bar{Q}_D$; and estimate the interceptor's required energy from $E/N_{01} = E_S/N_{01} + L_t + L_f$. It appears reasonable to assume that the two losses are additive even when, for example, L_f is smaller due to the filter bandwidth being greater than the signal bandwidth.

3.4 SYSTEMS THAT ALARM ON ANY THRESHOLD CROSSING

This section gives formulas and graphs for energy-detection systems having alarm circuits that simply threshold or that time-collapse and threshold. These are the systems labeled F-ED-T, FB-ED-T, and F-ED-C/T. The discussion on false-alarm rate also applies to the frequency-scanning system S-ED-T, but requires a small modification for the compressive system S/PC-ED-T. Frequency-scanning systems are covered separately in Section 3.6.

We continue to use the concept of dividing time-bandwidth space into cells of energy integration. (Figure 2.13), but sometimes with the following modifications.

(a) A grid model consisting of two overlapping grids is more appropriate when considering systems with overlapping filters, continuous energy integration, or overlapping I+D detection. (Note that there are four grids when filter and integration intervals both overlap.)

(b) A grid model is inappropriate for the F-ED-C/T system unless we can visualize a grid having slots of width T_2 spaced T_{p2} apart, where T_{p2} is the *pulse repetition interval* (PRI). Two kinds of integration times are involved: the single-pulse integration time T_2 and the system integration time $T \times$ (number of taps). The effective size of each time-bandwidth cell is $W \times$ (system integration time), and the area of the cell (usually $b_2T_2W_2$ or $b_3T_3W_3$) must be used instead of TW on the abscissa of the energy-detection graphs. Simple rules for calculations for time-collapsing systems are given in Section 3.4.6.

3.4.1 False-Alarm Rate

The FAR for the single-filter system F-ED-T is the product of the decision rate and the false-alarm probability Q_F. For I+D detection with integration time T, the decision rate is $1/T$; for overlapping I+D detection, the decision rate is $2/T$. As discussed in the previous section, overlapping I+D detection serves as a reasonable model for analyzing continuous-integration detection. The FAR for the filter-bank system FB-ED-T is FAR = (number of filters) $\times$ (decision rate per filter) $\times$ Q_F if the signal-present decisions that occur from different filters (filter-detector-threshold units) at the same time are treated as separate alarms. If a pair of simultaneous alarms from two overlapping filters elicits the same response from the operator as a single alarm, the true FAR is slightly smaller. The overlapping-filter case is the frequency-domain analog of the overlapping I+D detection case, and both are assumed to double the FAR. The decision rate per filter is that for the corresponding single-filter system given just above. The resulting formulas are summarized in Table 3.2, where the quantity $F = Q_F/TW$ is defined. Later, in Graph Set II, the FAR is held constant with changing TW by fixing the value of F.

The response of the intercept system's recognition stage to an alarm from the detection system must be taken into account for correctly defining the effective FAR. For example, if the action taken in the recognition stage is the same for simultaneous

Table 3.2 False-Alarm Rate for Systems that Alarm When Any Energy-Detector Threshold Crossing Occurs (The F-ED-T formulas can be used for the time-collapsing systems F-ED-C/T by using the system integration time in place of T. See Sec. 3.4.6.)

	False-Alarm Rate (FAR)		
Type of Energy Detection	*Single-Filter System F-ED-T*	*Filter-Bank System FB-ED-T*	*In terms of* $F = Q_F/TW$
I + D	Q_F/T	NQ_F/T	NWF
Continuous integration or overlapping I + D	$2Q_F/T$	$2NQ_F/T$	$2NWF$

FAR = False-alarm rate in alarms/second
Q_F = False-alarm probability at any decision time, or false-alarm probability per time-bandwidth cell of integration for I + D or overlapping I + D
T = Energy-detector integration time, seconds
N = Number of filter/detector units
W = Filter bandwidth, hertz

alarms as it is for a single one, the FAR is given (in the I+D case, with N non-overlapping filters) by

$$\text{FAR} = \frac{1 - (1 - Q_F)^N}{T} \tag{3.27}$$

This system is modeled by following an FB-ED-T system with an OR-gate. When $NQ_F \ll 1$, which should always be the case, then this formula and the one in Table 3.2 give very nearly equal results. If alarms occurring close together in time are treated as a single alarm by the recognition stage, then the FAR is lower.

3.4.2 Integration over Entire Transmission

We first consider an ideal single-filter energy-detection system (F-ED-T) with bandwidth W and integration time T, operating in a continuous-integration mode (Section 2.1.1). The acceptance band of width W covers the signal band of a transmission of bandwidth W_1, with $W \geq W_1$, and the integration time T is not less than the duration T_1 of the transmission. The time T_1 can be the look time for a scanning antenna or some function of the look time and the transmission duration (as described later in Section 3.7). There will be an instant (if $T = T_1$) or an interval (if $T > T_1$) of time when the integrated SNR will include the entire received signal energy E_1 of the transmission. Because noise peaks just before or just after the signal can cause a thresholding crossing, the probability P_d of detecting the transmission (in white Gaussian noise) is slightly greater than the probability Q_D calculated for a single energy integration (the Q_D given by equations and graphs throughout this chapter). The difference is usually small enough to be disregarded.

If the energy-detection system (F-ED-T) is operating in an I+D mode (Section 2.1.1) with the same system parameters and against the same transmission as described for the previous case, then the detection probability P_d is given by Q_D. Q_D is computed with $E_S = E_1$ only if the integration interval happens to coincide with the interval during which the signal is present at the detector input. In general, the detection probability is somewhat less than this because of the loss caused by splitting the signal in time. The exact expression for P_d is given by (3.19), with $P_d = \bar{Q}_D(E_1)$, but an estimate of the required E_1/N_{01} for a specified P_d can be found by letting $Q_D = P_d$, finding graphically the corresponding E_S/N_{01} (in dB), and letting $E_1/N_{01} = E_S/N_{01} + L_t$, where L_t is the splitting loss given in Section 3.3.3.

The above results for detection probability, and earlier results for FAR (Table 3.2), are summarized in Table 3.3. Unless there is a good reason not to, the computations of Q_F and Q_D should be made for $T = T_1$ and $W = W_1$. Estimates of Q_D for fixed Q_F can be found by using Figure 3.1 or a figure in Graph Set I, the choice depending on the parameter values considered. Graph Set II, given later, is useful if the FAR is fixed while various values of T are considered.

Table 3.3 Detection Formulas for Integration Over Entire Transmission (F-ED-T system with $T \geq T_1$ and $W \geq W_1$)

Type of Energy Detection	*False-Alarm Rate*	*Detection Probability*	*Q_D Corresponding to E_S =*
I + D	FAR = $Q_F/T = WF$	$P_d = Q_D$	$E_S = E_1 - L_t$ (in dB) L_t in Section 3.3.3
Continuous integration or overlapping I + D	FAR $\approx 2Q_F/T$	$P_d = Q_D$	$E_S = E_1$

$F = Q_F/TW$

E_1 = Interceptor's received energy from transmission.

3.4.3 Limited Integration Time

If T_1W_1 is so large that it is unreasonable to expect an interceptor to integrate the energy of an entire transmission (see Figure 3.19), computations can be made on the basis of *repeated observations*. Compute, or find graphically, Q_D for $E_S = E_1T_1/T$, and let

$$P_d \approx 1 - (1 - Q_D)^{T_1/T} \tag{3.28}$$

The FAR (for I+D) is Q_F/T, as given in Table 3.2. If T_1/T is large, the F-ED-T system will not perform so well as the F-ED-BMW system.

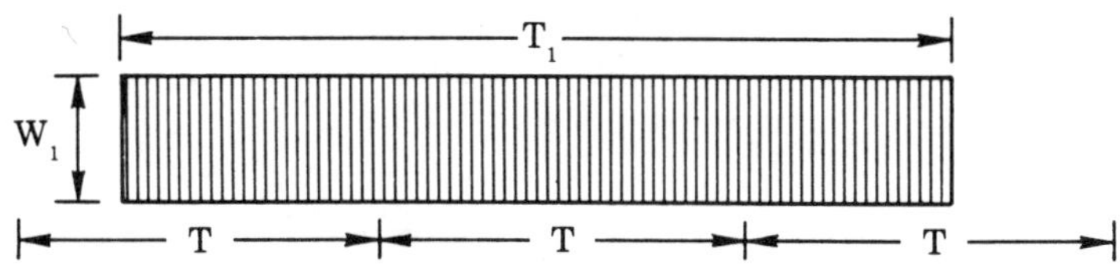

Figure 3.19 Integration time exceeded by transmission time.

3.4.4 Energy Integration Over Each Data Symbol

In the detection problem considered here, b_2 data symbols are transmitted, each as a signal occupying (not necessarily with uniform signal density) a rectangle $T_2 \times W_2$ in time-frequency space. The data symbols can be time-hopped or frequency-hopped, or be distributed in some other manner in a much larger rectangle of size $T_1 \times W_1$. The data symbols can be simple pulses ($T_2W_2 \approx 1$), groups of several pulses, or can have a pseudorandom microstructure ($T_2W_2 \gg 1$). The latter are sometimes referred to as spread-spectrum pulses, pseudonoise (PN) pulses, or direct-sequence (DS) signals. For now, we assume I+D energy detection.

By using the grid concept (Figure 2.13 and Section 3.2), we have b_2 cells containing signal energy and M_2 noise-only cells in the period of length T_1, where

$$M_2 = N\frac{T_1}{T_2} - b_2 \tag{3.29}$$

(Assume for now that the filters do not overlap in frequency.) When a single-filter system is used against time-hopped data symbols, this becomes

$$M_2 = b_2\left[\frac{T_{p2}}{T_2} - 1\right] \tag{3.30}$$

where T_{p2} is the time-hop period.

The quantity M_2 is introduced here because in many detection situations the fact that false alarms occur at fortuitous times should enter into the formula for the detection probability. If a signal decision in one of the M_2 empty time-bandwidth cells will lead to an examination of the message in a recognition stage as does a detection of one of the b_2 signals, then the detection probability (for the F-ED-T and FB-ED-T systems) should be defined as the probability of a threshold crossing during the message arrival time. This probability is (for an ideal grid fit)

$$P_d = 1 - (1 - Q_D)^{b_2}(1 - Q_F)^{M_2} \tag{3.31}$$

where Q_D is given by (3.5) with $E_S = E_2$, and E_2 is the interceptor's received energy per data symbol. In most cases of interest, the probability of a false alarm contributes an insignificant amount to the detection probability, and it is safe to use the approximation (plotted in Figure 3.20)

$$P_d \approx 1 - (1 - Q_D)^{b_2} \tag{3.32}$$

Note, however, that if Q_D is close to Q_F, (3.31) gives $P_d \approx b_2M_2Q_F$ while (3.32) gives $P_d \approx b_2Q_F$, and these can differ by many orders of magnitude.

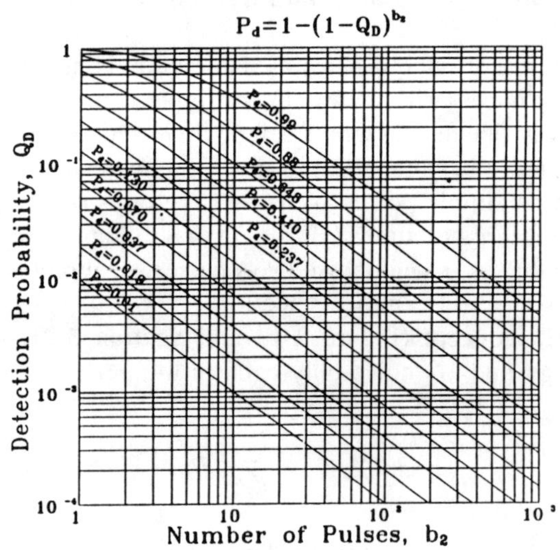

Figure 3.20 Per-pulse detection probability required to detect at least one of b_2 pulses with probability P_d.

In practice, false alarms outside the message period may also lead to a detection of the transmission. For example, suppose an alarm at any time t automatically results in an examination, in the recognition stage, of the receiver input voltage recorded over the interval of time $(t - T_1, t + T_1)$. Then there is a period of approximately $2T_1$ seconds during which an alarm will lead to an examination of a substantial portion of the message in the recognition stage. In this case, we would use $M_2 + NT_1/T_2$ in place of M_2 for computing the detection probability from (3.31). In a sometimes more convenient form, this probability is

$$P_d \approx 1 - (1 - Q_D)^{b_2}(1 - 2T_1\text{FAR}) \tag{3.33}$$

Equation (3.31) is usually the best to use for comparison purposes, but fortunately all three formulas give the same answer in most cases.

Equation (3.31) is based on I+D energy detection and the assumption that each data symbol is contained entirely in some time-bandwidth cell of integration. The loss L_f due to the signal being split randomly in frequency into different passbands is about the same for both I+D and continuous integration, and can be taken into account when estimating the received energy E_2. An additional loss L_t occurs for I+D detection when the signal is split randomly into two integration intervals. Losses L_f and L_t are discussed in Section 3.3.3.

The grid model for continuous integration and overlapping I+D has twice as many integration cells as does the I+D grid model. A clean model with no partly filled cells is impossible in this case, but (3.31) should be approximately valid for $M_2 = 2N(T_1/T_2) - b_2$.

Table 3.4 summarizes the detection formulas for integration over data symbols. Estimates of Q_D for fixed Q_F can be found by using Figure 3.1 or a figure in Graph Set I, the choice depending on the parameter values considered. Graph Sets II and III (Sections 3.4.7 and 3.4.8) are useful if the FAR is fixed while various values of T_2 are considered. Most applications will involve finding E_2 for a specified value of Q_D.

3.4.5 Energy Detection of Individual Elements

The analysis in Section 3.4.4 also applies to energy integration over individual elements, if an appropriate change in notation is made. Substitute $b_2 \times b_3$, T_3, W_3, E_3, and M_3 for b_2, T_2, W_2, E_2, and M_2, respectively, in the equations given in Section 3.4.4. Note that the total number of elements in the transmission is $b_2 \times b_3$ and that $E_3 = E_2/b_3$.

Table 3.4 Detection Formulas for Integration over Each Data Symbol [F-ED-T system: $N = 1$, $W = W_1$, $T = T_2$. FB-ED-T system: $W = W_2$, $T = T_2$, $N = W_1/W_2$ (or $2W_1/W_2 - 1$ if filters overlap).]

Type of Energy Detection	*False-Alarm Rate*	*Detection Probability*	*Number of Empty Integration Cells*	Q_D *Corresponding To* E_S =
Integrate and Dump (I + D)	$\text{FAR} = \frac{NQ_F}{T_2} = NWF$	$P_d = 1 - (1 - Q_D)^{b2}(1 - Q_F)^{M2} \approx 1 - (1 - Q_D)^{b2}$ for $M_2Q_F \ll b_2Q_D$	$M_2 = \frac{NT_1}{T_2} - b_2$	F-ED-T: $E_2 - L_t$ FB-ED-T: $E_2 - L_t - L_f$
Continuous integration or overlapping I + D	$\text{FAR} = \frac{2NQ_F}{T_2} = 2NWF$		$M_2 = \frac{2NT_1}{T_2} - b_2$	F-ED-T: E_2 FB-ED-T: $E_2 - L_f$

N = Number of filter/detector units

$F = \frac{Q_F}{TW} = \frac{Q_F}{T_2W}$

b_2 = Number of data symbols

E_2 = Received energy per data symbol

Above quantities are in decibel form. L_t and L_f are given in Section 3.3.

3.4.6 Time Collapsing of Periodic Signals

For the time-collapsing systems F-ED-C/T (Figure 2.11 and Table 2.3), the system integration time is given by

$$T' = (\text{number of taps}) \times T \tag{3.34}$$

The FAR (assuming I+D) is Q_F/T'. Substitute T' for T in the formulas for Q_D, and replace the abscissa TW with $T'W$ in the graph sets.

When data symbols are collapsed, then $T' = b_2T_2$, and the abscissa TW in graphs should be relabeled b_2T_2W. If the elements are in slot positions in each hop interval throughout the transmission and the elements rather than the data symbols are collapsed, then the parameter substitution given in Section 3.4.5 also applies here (giving $T' = b_2b_3T_3$). If element periodicity is only within a data symbol, then $T' = b_3T_3$.

3.4.7 Graph Set II

In Graph Set I, the false-alarm probability Q_F is held fixed in each figure. In Graph Set II (Figures 3.21–3.24), the FAR is held constant; therefore, Q_F varies with TW. Recall (Section 3.4.1 and Table 3.2, for I+D) that FAR $= NQ_F/T = NWQ_F/TW$, where $N = 1$ for the F-ED-T system and where $N = W_1/W$ (nonoverlapping filters) or $N = 2W_1/W - 1$ (overlapping) for the FB-ED-T system. By holding $F = Q_F/TW$ constant, FAR is constant for fixed NW, that is, for fixed W_1.

GRAPH SET II

Detection probability Q_d per cell containing signal *versus* time-bandwidth product TW of cells for constant factor F of false-alarm rate (FAR) with curves given for values of E_S/N_{01}:

FAR	$= NWF$ (alarms/sec)
	$= W_1F$ for $N = W_1/W$ nonoverlapping filters
F	$= Q_F/TW$
Q_F	= per-cell false-alarm probability
W_1	= Transmission bandwidth
E_S	= Signal energy in cell containing signal
N_{01}	= One-sided noise power spectral density

Figure	F
3.21	10^{-11}
3.22	10^{-10}
3.23	10^{-9}
3.24	10^{-8}

NOTES

1. For integration over the transmission with the F-ED-T system, let $TW = T_1W_1$, $E_S = E_1$, and $P_d = Q_d$.
2. For pulse detection of b_2 pulses with the F-ED-T or the FB-ED-T system, let $TW = T_2W_2$, $E_S = E_2$, and $P_d \approx 1 - (1 - Q_d)^{b_2}$.
3. To detect periodic pulses with the time collapsing system F-ED-C/T, change the abscissa label TW to $b_2T_2W_1$, and let $E_S = E_1$ and $P_d = Q_d$.
4. The expression for FAR assumes I+D energy detection. Continuous integration has approximately double this rate.

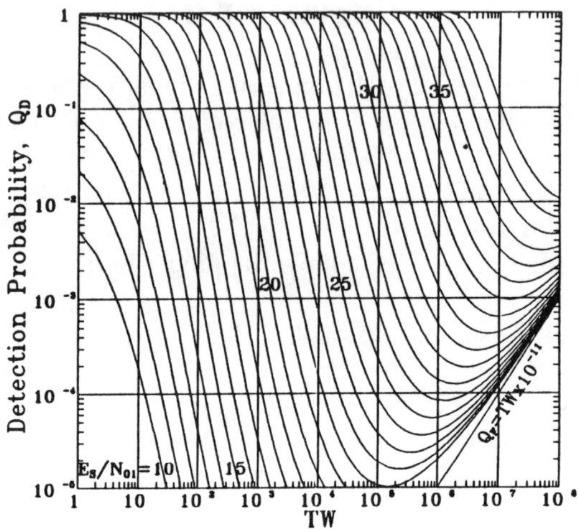

Figure 3.21 $F = 10^{-11}$ (FAR $= TWF$; generally, assume FAR $= W_1 \times 10^{-11}$).

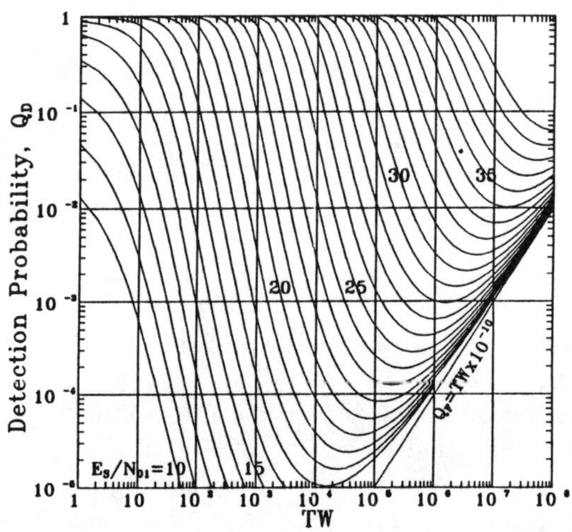

Figure 3.22 $F = 10^{-10}$ (FAR $= TWF$; generally, assume FAR $= W_1 \times 10^{-10}$).

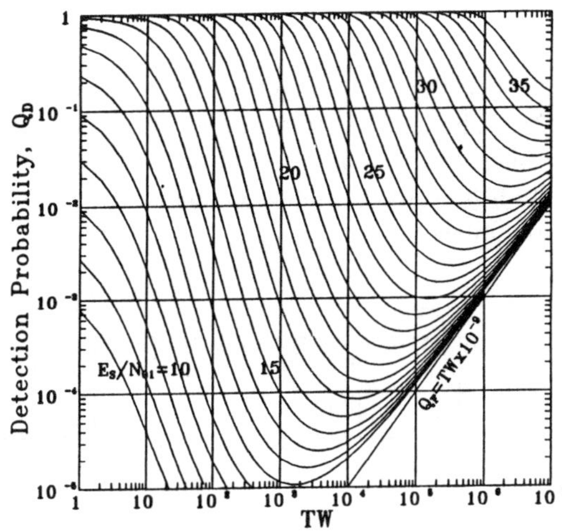

Figure 3.23 $F = 10^{-9}$ (FAR = TWF; generally, assume FAR = $W_1 \times 10^{-9}$).

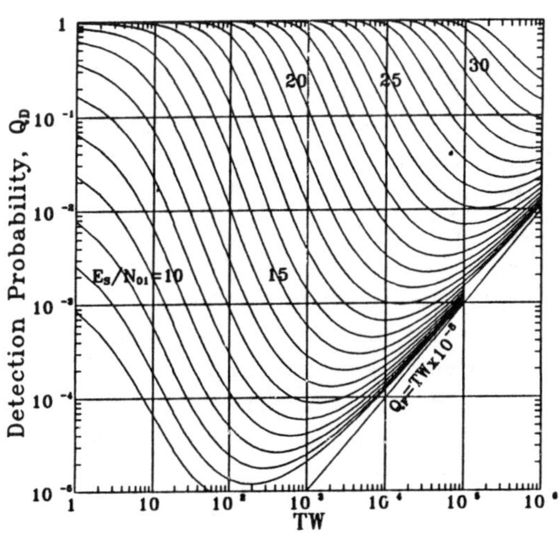

Figure 3.24 $F = 10^{-8}$ (FAR = TWF; generally, assume FAR = $W_1 \times 10^{-8}$).

Example 1

Signals A and B have the same bandwidth but have different durations, as indicated in Figure 3.25. Whichever is transmitted, the interceptor knows the signal band and duration. He integrates the energy over the entire signal, matching his filter bandwidth and integration time to the signal bandwidth and duration. Compare the detectability of the two signals.

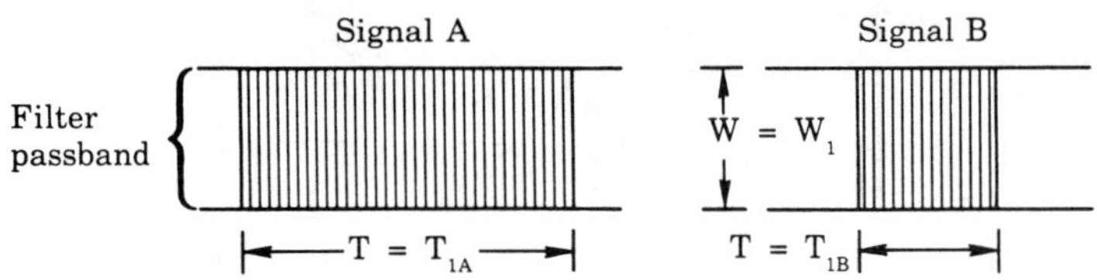

Figure 3.25 Comparing detectability of two signals having equal bandwidth but different durations.

Procedure: The appropriate formulas for this type of detection were given in Table 3.3. Select a graph in Graph Set II with a suitable value of FAR (doubling the stated value if continuous integration is assumed). In the use of this graph, E_S is the total signal energy that the interceptor receives from the signal (after taking into account a signal-splitting loss L_t if I+D energy detection is used). Because integration is over the entire signal, P_d is approximately equal to Q_D. From the selected graph, find the two values of $Q_D = P_d$ for the two values of $TW = T_1W_1$, using the same curve for both if the intercepted energy E_S is the same for signals A and B.

If Q_D is close to Q_F or is close to unity for either signal, the comparison of detection probabilities is quantitatively poor. In this case, alter the assumptions (e.g., assume a shorter intercept range) and find the new values of Q_D. This problem can usually be avoided by specifying the detection probability P_d and comparing other quantities. The first step is to find on the selected graph the respective values of the required E_S/N_{01} for a fixed $Q_D = P_d$. If the transmitted energy required for a specified communication performance is the same for signals A and B, then the difference (in dB) between the two values of E_S/N_{01} (adjusted for any differences in interceptor losses) is a good measure of the relative detectability of the two signals. This difference is about 4.2 $\log_{10}r$ for FAR $\approx W_1 \times 10^{-8}$ and is about 4.4 $\log_{10}r$ for FAR $\approx W_1 \times 10^{-11}$, where r is the ratio of the T_1 values and $T_1W_1 > 10^4$ and $Q_F \ll Q_D$. (The difference would be 5 $\log_{10}r$ if Q_F, rather than FAR, were held constant.) If the required transmitted energy is different for the two signals, the difference (in dB) in transmitted energy is subtracted from the difference of E_S/N_{01} values to find a measure of the relative detectability of the two signals.

Example 2

Systems A and B are filter-bank, energy-detection systems searching for the same frequency-hopped (and possibly time-hopped) transmission consisting of b_2 data symbols or pulses, each of size T_2W_2. Detection system A uses only a few filters, each of bandwidth $W \gg W_2$, to cover the transmission band, and system B uses many filters each of bandwidth $W = W_2$, as indicated in Figure 3.26. Compare the detection probabilities of the two systems, assuming that they use nonoverlapping filters and have the same FAR. Do this by estimating the energy each requires for a specified P_d.

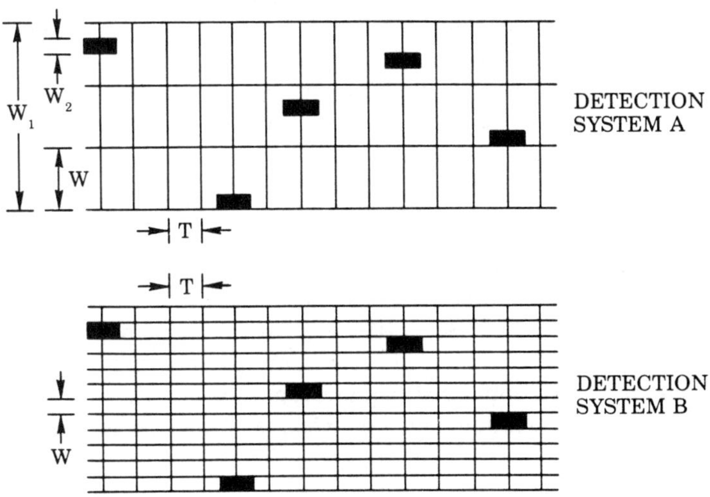

Figure 3.26 Filter-bank detection: $W \gg W_2$ (system A) and $W = W_2$ (system B).

Procedure: The formulas for simple filter-bank detection (with the FB-ED-T system) were given in Table 3.4. We assume that the integration time T of both systems is T_2. (However, either system can have a larger T, in which case T is substituted for T_2 in the equations in Table 3.4.) The number N of filters has a different value for the two systems, but their total frequency coverage NW is the same.

Select a graph with that value of NW and with an appropriate FAR (doubling the stated FAR if continuous integration is used). When possible, select a graph such that $2 \times \text{FAR} \times T_1 \ll P_d$, which is approximately equivalent to the condition ($M_2Q_F \ll b_2Q_D$) that enabled us to use the approximation $Q_D \approx 1 - (1 - P_d)^{1/b_2}$ obtained from (3.32). Otherwise, for a fixed value of P_d, Q_D will have a somewhat different value for the two systems, because M_2 is different for each.

Find on the selected graph the respective values of required E_S/N_{01} for the two values of TW. In this use of the graph, E_S represents the interceptor's *per-pulse* received energy minus any signal-splitting losses ($E_S/N_{01} = E_2/N_{01} - L_t - L_f$), all in dB. The loss L_f due to splitting the signal in frequency will be very small for system A, because $W \gg W_2$. The dB difference in the two values of E_2/N_{01} is a measure of the relative detection capabilities of the two systems.

Example 3

We now compare a single-filter, energy-detection system called system C (see Figure 3.27) with detection system A or B of Example 2. System C matches its filter bandwidth and integration time to the bandwidth and duration of the entire transmission as shown in Figure 3.27. To give systems A, B, and C the same FAR is not entirely fair because the amount of processing in a later recognition stage is different, but for convenience the same FAR is assumed for all three.

Procedure: Estimate the value of E_2/N_{01} required by system A or B by using the procedure described in Example 2. On the same graph, find the required E_S/N_{01} corresponding to $Q_D = P_d$. For system C, $E_S/N_{01} = E_1/N_{01} + L_t$ (in dB), where E_1 is the energy received from the entire transmission. Because $E_1 = b_2E_2$, we compare the quantity $E_2/N_{01} = E_S/N_{01} + L_t - 10\log_{10}b_2$ with the E_2/N_{01} required by system A or B. An additional processing gain of several dB should be given to system A or B if it is reasonable to assume that the interceptor would use additional processing such as a binary moving-window detector. Chapter 4 treats this in detail.

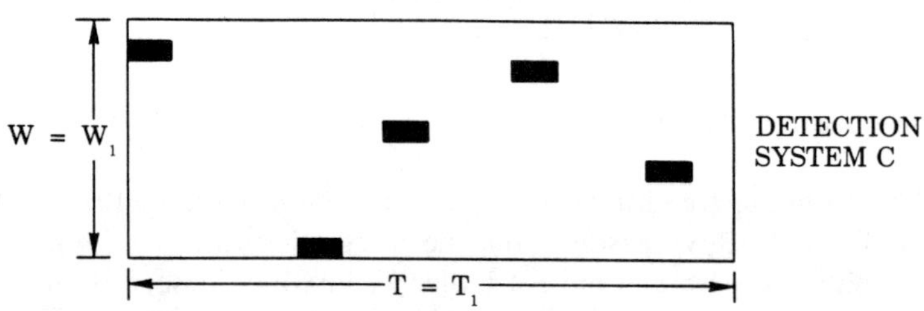

Figure 3.27 Energy integration over transmission: comparison with systems A and B of Figure 3.26.

3.4.8 Graph Set III

Graph Set III (Figures 3.28 to 3.34) can be used to compare a simple pulse-detection system (F-ED-T or FB-ED-T, with $T = T_2$) with a system that integrates energy over the entire transmission (F-ED-T with $T = T_1$ and $W = W_1$). Instructions are given in Chapter 4 for also using these graphs to compare the F-ED-BMW and FB-ED-

OR/BMW systems against integration over the entire transmission.

When making comparisons, remember that E_1 is the total energy from the transmission received by the interceptor ($E_1 = b_2E_2$). Filter-bank systems with nonoverlapping filters can suffer a frequency-splitting loss L_f (Section 3.3.3) not taken into account in these graphs.

In Figure 3.28, b_2 pulses of size $T_2W_2 = 1$ are frequency-hopped or time-hopped in a transmission of size T_1W_1. Curves on the left are for simple pulse detection. Curves on the right are for *integration over the transmission* (IOT).

Two detection systems or situations being compared in Figures 3.29 to 3.34 can involve different values of T, W, and N, and still have the same FAR on any given graph, provided NW is the same for both systems. This is because $F = Q_F/TW$ is held constant in each graph while $TW = T_1W_1$ and $TW = T_2W$ vary, respectively, for IOT and for pulse detection. The abscissa in these figures is labeled T_2W_2 but can be interpreted as T_2W. An F-ED-T system with $W = W_1$ is sometimes a better pulse-detection system than the FB-ED-T system with $W = W_2$, in which case the abscissa can be read as T_2W_1.

Because F is fixed and Q_F is always less than Q_D, we have the restriction $TWF < Q_D$. This restriction becomes $T_1W_1 < P_d/F$ when $T = T_1$ and $T_2W < Q_D/F$ when $T = T_2$; this explains the difference in parameter coverage for the various graphs.

GRAPH SET III

Comparisons of interceptor's required E_1/N_{01} for integration over the transmission and simple pulse detection.

Integration over the transmission (IOT):

E_1 = Total energy of transmission; $P_d = Q_d$
F-ED-T system with $T = T_1$ and $W = W_1$
Figure 3.28: Use graph on right.
Figures 3.29–3.32: Upper abscissa gives T_1W_1.

Draw a horizontal line through the point on the IOT curve corresponding to the value of T_1W_1 on the upper abscissa. Note intersection(s) with curve(s) for simple pulse detection.

Simple Pulse Detection:

b_2 = Number of pulses
$E_2 = E_1/b_2$ = Per-pulse energy; $P_d = 1 - (1 - Q_d)^{b_2}$
F-ED-T system with $T = T_2$ and $W = W_1 = W_2$
FB-ED-T system with $T = T_2$, $W = W_2$, and N filters
Figure 3.28 ($T_2W_2 = 1$): Use graph on left.
Figures 3.29–3.34: Lower abscissa gives T_2W_2.

Figure	T_2W	$F = Q_F/TW$	P_d
3.28	1	$10^{-11}, 10^{-8}$	0.1,0.5,0.9
3.29	1--10^5	10^{-11}	0.1
3.30	1–10^5	10^{-11}	0.5
3.31	1–10^5	10^{-11}	0.9
3.32	1–10^5	10^{-10}	0.5
3.33	1–10^4	10^{-9}	0.5
3.34	1–10^3	10^{-8}	0.5

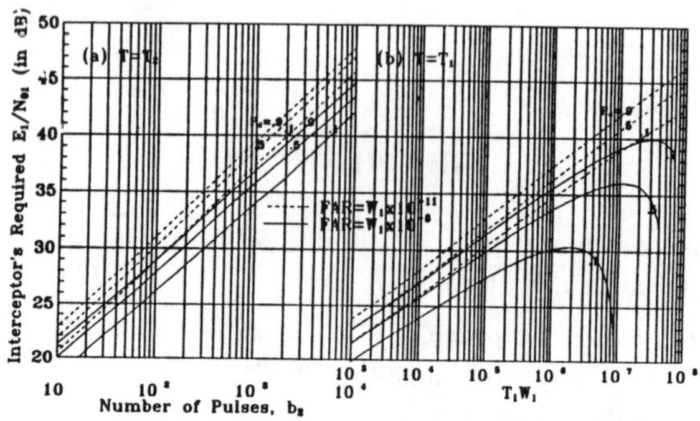

Figure 3.28 Detection of FH, TH, or FH/TH signals ($T_2W_2 = 1$): (a) pulse detection with an F-ED-T system ($T = T_2$, $W = W_2 = W_1$) or FB-ED-T system ($T = T_2$, $W = W_2 = W_1/N$); (b) energy integration over transmission with an F-ED-T system ($T = T_1$, $W = W_1$).

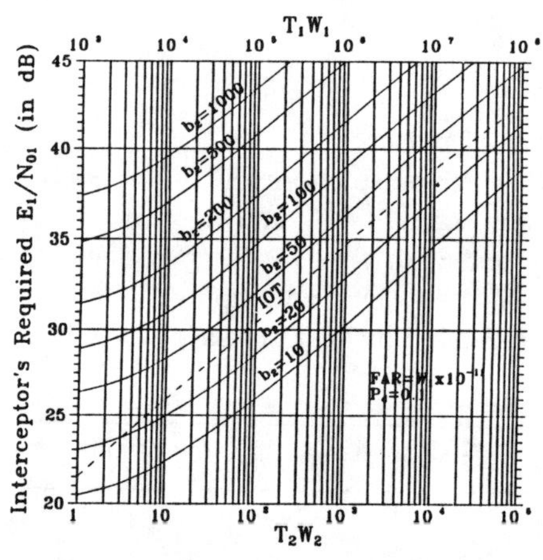

Figure 3.29 Detection curves for comparing energy integration over transmission (IOT) with integration over individual pulses. Dashed curve (upper abscissa): IOT. Solid curves (lower abscissa): simple pulse detection.

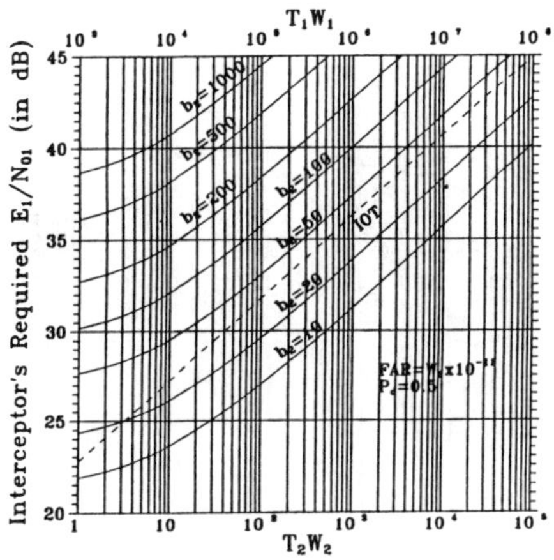

Figure 3.30 Detection curves for comparing energy integration over transmission (IOT) with integration over individual pulses. Dashed curve (upper abscissa): IOT. Solid curves (lower abscissa): simple pulse detection.

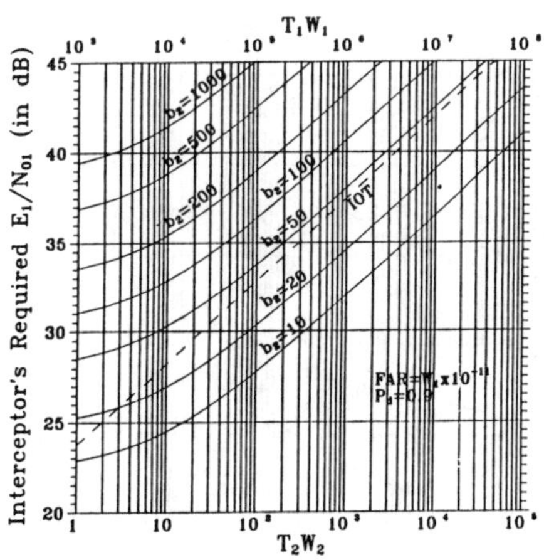

Figure 3.31 Detection curves for comparing energy integration over transmission (IOT) with integration over individual pulses. Dashed curve (upper abscissa): IOT. Solid curves (lower abscissa): simple pulse detection.

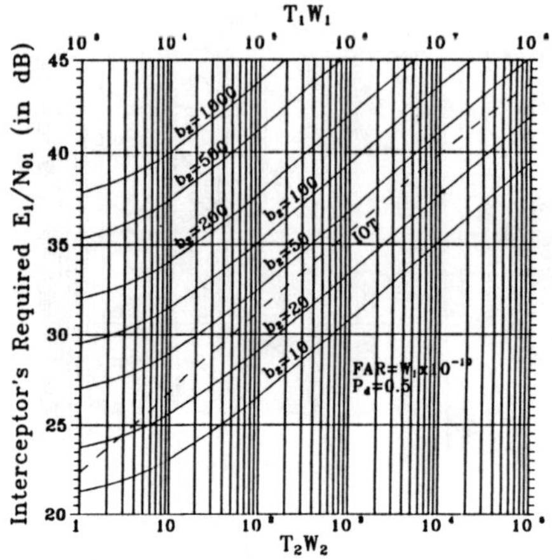

Figure 3.32 Detection curves for comparing energy integration over transmission (IOT) with integration over individual pulses. Dashed curve (upper abscissa): IOT. Solid curves (lower abscissa): simple pulse detection.

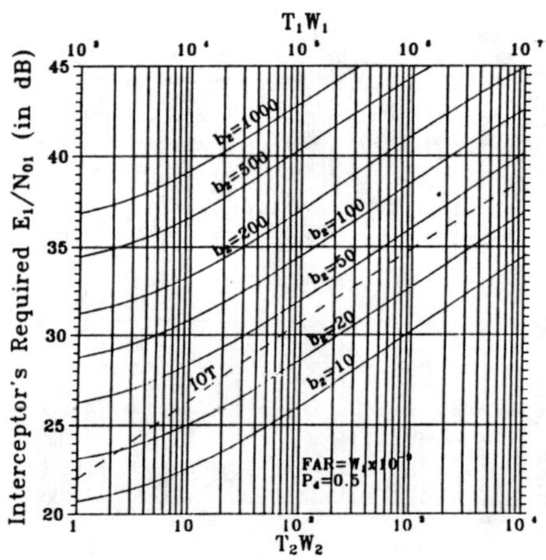

Figure 3.33 Detection curves for comparing energy integration over transmission (IOT) with integration over individual pulses. Dashed curve (upper abscissa): IOT. Solid curves (lower abscissa): simple pulse detection.

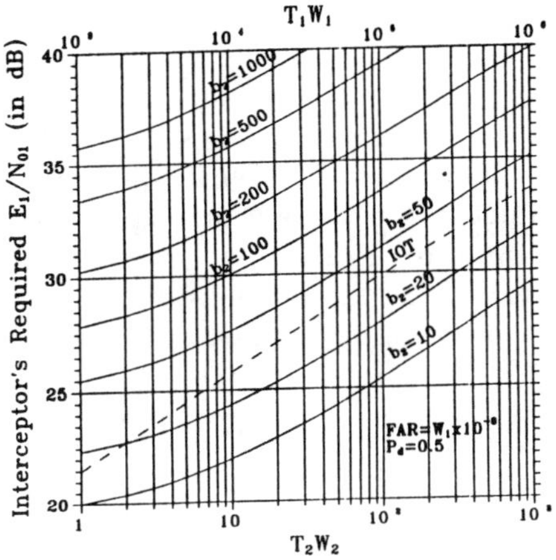

Figure 3.34 Detection curves for comparing energy integration over transmission (IOT) with integration over individual pulses. Dashed curve (upper abscissa): IOT. Solid curves (lower abscissa): simple pulse detection.

3.5 BINARY MOVING-WINDOW DETECTION

The two kinds of *binary moving-window detector* (BMWD) systems most likely to be used are the F-ED-BMW and the FB-ED-OR/BMW systems. The F-ED-BMW system (Figure 2.12 and Table 2.4) is a single-filter system for detecting time-hopped signals. The output from the thresholder at time jT is $y_j = 0$ or 1. To simplify computations for the F-ED-BMW system, a system (F-ED-OR/BMW) synchronized in time with the time-hop intervals is hypothesized. The FB-ED-OR/BMW system (Figure 2.15 and Table 2.7) is a filter-bank system for detecting frequency-hopped signals with or without time hopping. The output from the OR-gate at time jT is $y_j = 0$ or 1. A synchronized model of the FB-ED-OR/BMW system is hypothesized to simplify computations.

The formulas given here are for the $T = T_2$ case (energy integration over data symbols), but instructions are given later for using them in the $T = T_3$ case (integration over elements). The grid model described in Section 3.2 is applicable here. The size of time-bandwidth cells of energy integration is T_2W, where $W = W_1$ for the F-ED-BMW system and $W = W_2$ for the FB-ED-OR/BMW system. In the signal case, b_2 of the cells with signal will contain signal and all remaining cells are "empty." The probability of a signal-present decision (i.e., an output 1 from the thresholder) is Q_F in an empty cell and is Q_D in a cell with signal. The formulas are based on I+D energy detection and assume that each data symbol is contained entirely in some integration cell. Losses resulting from the signal being split into two integration intervals or into different frequency passbands are explained in Section 3.3.3.

The probability at any time jT_2 that a 1 enters the BMWD is denoted by p_0 when only noise is present during the previous T_2 seconds, and is denoted by p_1 when a signal is present (in one of the N passbands). That is,

$$p_0 = P[y_j = 1|\text{noise}] \tag{3.35}$$

and

$$p_1 = P[y_j = 1|\text{signal}] \tag{3.36}$$

For the F-ED-BMW systems, the formulas for p_0 and p_1 are simply

$$p_0 = Q_F \tag{3.37}$$

and

$$p_1 = Q_D \tag{3.38}$$

For the FB-ED-OR/BMW system, the formulas are

$$\begin{aligned} p_0 &= 1 - (1 - Q_F)^N \\ &\approx NQ_F \quad \text{for} \quad NQ_F \ll 1, \end{aligned} \tag{3.39}$$

and

$$\begin{aligned} p_1 &= 1 - (1 - Q_F)^{N-1}(1 - Q_D) \\ &\approx Q_D \quad \text{for} \quad NQ_F \ll Q_D \ll 1 \end{aligned} \tag{3.40}$$

At each time jT_2, the BMWD computes

$$S_j = \sum_{k=j-M+1}^{j} y_k \tag{3.41}$$

and compares it with the threshold K_M, where $M = T_1/T_2$ is the window length. The system alarms when the sum S_j crosses the threshold K_M in an upward direction; that is, when $S_{j-1} = K_M - 1$ and $S_j = K_M$. The probability of this joint event when no signal is present is

$$P_T = \binom{M-1}{K_M-1} p_0^{K_M}(1-p_0)^{M-K_M+1} \tag{3.42}$$

This expression and a reference to its derivation are used by Hansen [18]. It is the product of three component probabilities: $P_T = P\{y_{j-M} = 0\} \cdot P\{y_{j-M+1} + \ldots + y_{j-1} = K_m - 1\} \cdot P\{y_j = 1\}$. The FAR, in alarms per seconds, is

$$\text{FAR} = P_T/T_2 \tag{3.43}$$

The probability of a detection by the BMWD is approximately equal to (and bounded below by) the probability that there are K_M 1s in the window when the window is aligned in time with the transmission. This probability is given by

$$P_d = \sum_{i=K_M}^{M} \sum_{j=\max(0,i-M+b2)}^{\min(i,b_2)} \binom{b_2}{j}\binom{M-b_2}{i-j} p_1^j p_0^{i-j} \cdot (1-p_1)^{b_2-j}(1-p_0)^{M-i+j-b_2} \tag{3.44}$$

When computing P_d using (3.44), note that the summand can be computed iteratively as j is increased by 1:

$$f(j) = f(j-1)\left[\left(\frac{b_2-j+1}{j}\right)\left(\frac{i-j+1}{M-i+j-b_2}\right)\left(\frac{p_1}{p_0}\right)\left(\frac{1-p_0}{1-p_1}\right)\right] \tag{3.45}$$

where $f(j)$ is the summand in (3.44).

An exact procedure for calculating P_d for the BMWD is described by Dillard [19]. For frequency hopping alone, with $M = b_2$, (3.44) becomes

$$P_d = \sum_{i=K_M}^{b_2} \binom{b_2}{i} p_1^i (1-p_1)^{b_2-i} \tag{3.46}$$

This equation also applies to an F-ED-OR/BMW system synchronized with a TH signal (Table 2.4), where

$$p_1 = 1-(1-Q_F)^{T_{p2}/T_2-1}(1-Q_D), \tag{3.47}$$

and to the synchronized FB-ED-OR/BMW system (Table 2.7), where

$$p_1 = 1 - (1 - Q_F)^{NT_{p2}/T_2 - 1}(1 - Q_D) \tag{3.48}$$

For the F-ED-BMW system and the unsynchronized FB-ED-OR/BMW system, the use of (3.47) or (3.48), respectively, serves as a lower bound and approximation to (3.44). The window length used in computing P_T, however, remains $M = bT_{p2}/T_2$.

A useful relationship to note when computing P_T for $K_M = 1, 2, \ldots$, is

$$\binom{M-1}{K_M-1} = \binom{M-1}{K_M-2}\left[\frac{M-K_M+1}{K_M-1}\right] \tag{3.49}$$

Usually, FAR or P_T is specified and p_0 must be computed. Successive approximations, starting with $p_0^{(i)} = 0$, can be made by using the relationship [15]:

$$p_0^{(i+1)} = C_p(1 - p_0^{(i)})^{-J/K_M} \tag{3.50}$$

where

$$C_p = \left[\frac{P_T}{\binom{M-1}{K_M-1}}\right]^{1/K_M} \tag{3.51}$$

and $J = M - K_M + 1$. The number of iterations required increases as K_M/M increases. For some values of P_T, M, and K_M, there is no solution. By differentiating the right-hand side of (3.42) with respect to p_0 and equating the result to zero, we see that the maximum value occurs whenever $p_0 = K_M/(M + 1)$. If P_T exceeds this maximum then there is no solution; otherwise, the solution of interest is the one for which $p_0 \le K_M/(M + 1)$. When $M - K_M \ge 10$ and $K_M \ge 10$, Stirling's approximation to factorials can be used, giving

$$C_p = \frac{K_M}{M}\left[\frac{MP_T}{e}\sqrt{\frac{2\pi M}{JK_M}}\left(\frac{1 + (M + 1 + 1/12)/12JK_M}{1 + 1/12M}\right)\left(\frac{J}{M}\right)^{J}\right]^{1/K_M} \tag{3.52}$$

For the F-ED-BMW system, $Q_F = p_0$. For the FB-ED-OR/BMW system, $Q_F = 1 - (1 - p_0)^{1/N}$. In calculations for an F-ED-OR/BMW system modified to synchronize the OR function over the hop interval for a TH transmission, the window length is $M = b_2$ instead of $M = T_1/T_2$, (3.43) becomes FAR $= P_T/T_{p2}$, and

$$Q_F = 1 - (1 - p_0)^{T_2/NT_{p2}} \tag{3.53}$$

(This is exactly equivalent to substituting $N^* = NT_{p2}/T_2$ for N and $T_2^* = T_{p2}$ for T_2

in previous equations for FH for the FB-ED-OR/BMW.) These relationships are valid also for the F-ED-OR/BMW system, with $N = 1$, where

$$Q_F = 1 - (1 - p_0)^{T_2/T_{p2}} \tag{3.54}$$

In each case, the resulting value of Q_F can be used in (3.2) to find the energy detection threshold K. (Note that $K = -2 \ln Q_F$ when $T_2W_2 = 1$, and $K_0 = K/2$ in Figure 3.2.) This value of K is used in computations of Q_D, such as when using (3.3). Alternatively, calculation of K can be bypassed by using graphs in Section 3.2.

Finding p_1 given a value of P_d is a more difficult numerical problem. A binomial table can be used for frequency-hopped cases where $M = b_2$. For large b_2, the normal distribution approximates the binomial distribution, giving $p_1 \approx K_M/b_2$ for $P_d = 0.5$. If $M > b_2$, one can either use a simple programmable calculator to find P_d over a range of values of p_1, then graphically interpolate for p_1, or use a computer to inversely find p_1 for a given P_d. When $K_M = 2$, the expression for P_d is not too unwieldy:

$$\begin{aligned} P_d = 1 - [1 + (b_2 - 1)p_1 + (M - 1 - b_2)p_0 - (M - 1)p_0p_1] \\ \cdot (1 - p_1)^{b_2-1}(1 - p_0)^{M-1-b_2} \end{aligned} \tag{3.55}$$

This leads to an expression that can be used for successive approximations of p_1, starting with $p_1^{(1)} = 1 - (1 - P_d)^{1/b_2}$:

$$p_1^{(i+1)} = 1 - \left[\frac{(1 - p_d)/(1 - p_0)^{M-1-b_2}}{p_1^{(i)}[b_2 - 1 - (M - 1)p_0] + 1 + (M - 1 - b_2)p_0}\right]^{1/(b_2-1)} \tag{3.56}$$

In many cases when M is larger than about twenty, letting $K_M = 2$ gives roughly half of the dB improvement that the optimum K_M has over $K_M = 1$. This can be seen in Figure 3.35, for some pure FH signal cases. Calculations for $K_M = 2$ are sometimes useful, then, for estimating the improvement possible by using the BMWD.

Figure 3.36 illustrates how the curve of required energy *versus* threshold K_M is fairly flat for a wide region of K_M around the optimum value. The optimum value of K_M for the F-ED-BMW and FB-ED-OR/BMW systems is a function of all parameters involved, but is almost always less than $0.5b_2$. The curve of required energy *versus* K_M usually begins to flatten at about $0.1b_2$ or $0.2b_2$ and often remains fairly flat well beyond $0.5b_2$.

To reduce the number of parameters that must be considered when computing detection performance for BMWD systems, we define and use in Chapter 4 the concept of *density*. The density D is the ratio of time-bandwidth area occupied by the signal to that of the whole transmission.

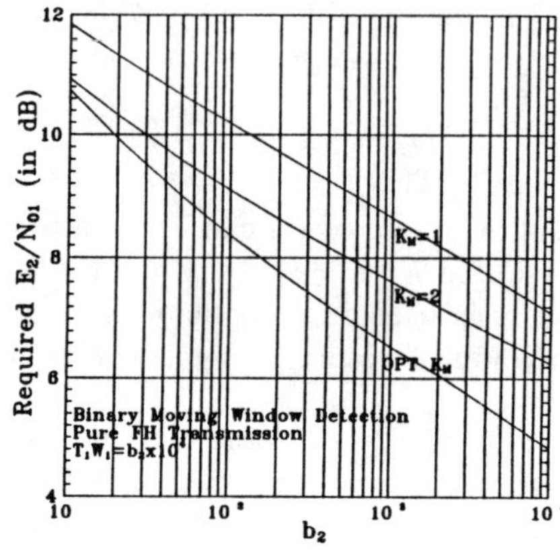

Figure 3.35 Required per-pulse energy-to-noise ratio *versus* number of pulses, for the optimum threshold (OPT K_M) and for $K_M = 1$ and $K_M = 2$; FB-ED-OR/BMW system, N = number of filters = $W_1/W_2 = 10^4$, $Q_D = 0.5$, FAR = $W_1 \times 10^{-11}$.

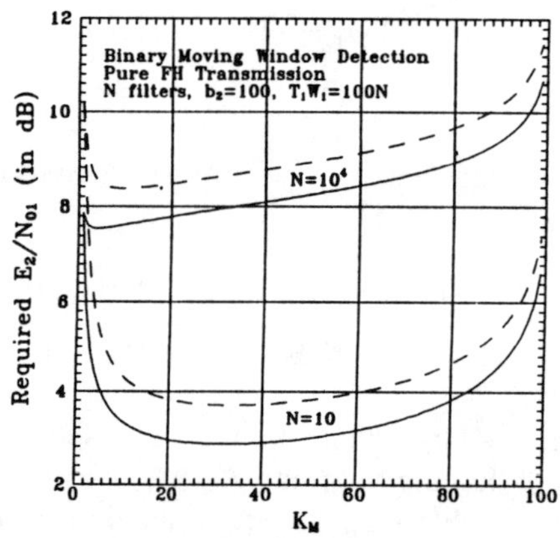

Figure 3.36 Illustration of performance as a function of K_M; FB-ED-OR/BMW system, N = number of filters = W_1/W_2, $Q_D = 0.5$. For dashed curves FAR = $W_1 \times 10^{-11} = NW_2 \times 10^{-11}$; for solid curves FAR = $W_1 \times 10^{-8} = NW_2 \times 10^{-8}$.

Instructions are given below for making computations for other detection systems (described in Sections 2.2.3 and 2.3.2) that involve BMWD.

Integration over Elements with the F-ED-BMW or FB-ED-OR/BMW System: The window length is $M = T_1/T_3$. (If a synchronized system is assumed, use FB-ED-OR/BMW equations with $N^* = T_{p3}/T_3$ for N, $T_3^* = T_{p3}$ for T_3, and $M = T_1/T_{p3}$.) Use T_3 in place of T_2 in the computation of Q_D and in (3.43). Use b_2b_3 in place of

b_2 in (3.44) or (3.46).

Integration over Elements with the FB-ED-bank/BMW System: (See Table 2.6.) Use F-ED-BMW equations with window length $M = T_2/T_3$. (Except, against TH elements, use FB-ED-OR/BMW equations with $N^* = T_{p3}/T_3$ for N, $T_3^* = T_{p3}$ for T_3, and $M = T_2/T_{p3}$.) Use b_3 for b_2 in (3.44) or (3.46) to find P_d per data symbol. The probability of detecting the transmission is $P_d' = 1 - (1 - P_d)^{b_2}$, where P_d is the detection probability per data symbol. Equation (3.43) becomes FAR $= NP_T/T_2$ (not replacing N with N^*).

Covering Possible Transmission Bands with the F-ED-bank/BMW System: (See Table 2.6.) For a bank of N F-ED-BMW systems, (3.43) becomes FAR $= NP_T/T_2$.

Binary Time-Collapsing of Periodic Frequency-Hopped Data Symbols with the FB-ED-OR/C/BMW System: Use window length $M = T_1/T_{p2} = b_2$. Other equations are also the same as for pure FH and for FH/PN. (See Figure 2.16 and Table 2.8.)

3.6 CALCULATIONS FOR FREQUENCY-SCANNING SYSTEMS

Two kinds of frequency-scanning receivers were described in Section 2.4, a conventional one and a compressive receiver. With either, a binary moving-window detector can be employed as an alarm circuit. The resulting four systems we consider are the S-ED-T system (Figure 2.19), the S-ED-BMW system (Figure 2.20), the S/PC-ED-T system (Figure 2.22), and the S/PC-ED-BMW system (Figure 2.23). Detecting a weak PN transmission requires sweeping across the transmission and integrating signal energy over the duration of the intersection. Detecting a hopped transmission requires intersecting with at least one pulse. If the system can intersect with more than a few pulses, the use of binary moving-window detection will increase the probability of detecting the transmission. Section 3.6.1 discusses the detection of individual pulses. Section 3.6.2 briefly discusses the use of formulas from Section 3.5 for binary moving-window detection. Section 3.6.3 addresses the situation in which the data symbols consist of hopped pulses. Section 3.6.4 considers the detection of PN transmissions. Other methods of calculating detectability, applicable mainly to radar pulses, can be found in Wiley [20].

Before proceeding with calculations, we remind the reader that vulnerability to detection should be based on frequency-scanning receivers only when they are known to be the only threat or are being considered as a secondary threat. An ideal filter-bank detection system with parameters matched to those of the signal will always perform better than an ideal frequency-scanning detection system with an equivalent alarm circuit because a frequency-scanning system uses only a part of the available signal information.

3.6.1 Frequency Scanning over Spread-Spectrum Pulses

A *coincidence* is an intersection, in time and frequency, of the receiver sweep path with a pulse. Formulas are given here for estimating the interceptor's coincidence probabilities, both for the case where the intercepted energy is greater than half that of a spread-spectrum pulse and where the energy is less than half.

In the situation of interest here, the interceptor's SNR will be small enough that he must sweep over a significant portion of a spread-spectrum pulse to accomplish energy detection. We assume that the energy of each spread-spectrum pulse is spread uniformly over its time-bandwidth area and that the interceptor's receiver has a rectangular-shaped IF passband. (See Section 2.2.)

The slope of the receiver sweep path (frequency/time) is s, the pulse duration is T_2, and the pulse bandwidth is W_2. If the interceptor's bandwidth W is less than $W_2 + T_2s$, it is impossible for him to accept the entire energy of the pulse; if his bandwidth W is greater than $(W_2 + T_2s)/2$, the center of his sweep path can vary in frequency as much as W and he can still accept at least half the signal energy. This leads us to express the receiver bandwidth as

$$W = r(W_2 + T_2s) \tag{3.57}$$

Calculations of coincidence probabilities are very much simplified if the restriction $r \geq 1/2$ is imposed; that is, if $W \geq (W_2 + T_2s)/2$. This restriction is justified when considering energy detection of weak signals.

We denote by p' the probability of accepting some portion less than half the energy of a pulse, and by p'' the probability of accepting at least half. By inspecting Figures 2.21 or 2.24, we see that the probability of intersecting with some portion of the pulse is

$$p' + p'' = \frac{W + W_2 + sT_2}{W_0} \tag{3.58}$$

Compressive receivers can operate at very high scan rates, so we should note that $p' + p'' = 1$ if $T_2 + W_2/s > W_s/s$. That is, the receiver is certain to accept energy from a pulse that is too large to fit between two consecutive sweep paths. For $r \geq 1/2$, the single-pulse coincidence probabilities are

$$p'' = \frac{r(W_2 + sT_2)}{W_0} \tag{3.59}$$

$$p' = \begin{cases} p''/r & \text{if} \quad W_s > W_2 + sT_2 \\ 1 - p'' & \text{if} \quad W_s \leq W_2 + sT_2 \end{cases} \tag{3.60}$$

The probability $p' + p''$ that some portion of the energy of a particular pulse is accepted is sometimes called the *unit coincidence probability*.

The interceptor's choice of a sweep slope s can depend on such factors as pulse dimensions, transmission duration, time-hop period, hardware limitations, and compatibility with other signal-intercept operations. In the discussions below, we assume that the pulse dimensions (T_2, W_2) are fixed and known by the interceptor and that a sweep rate has been selected. The selection of sweep rate is discussed later.

Computation of the interceptor's detection probability should use values of the receiver bandwidth W and energy-integration time T that are at least close to optimum, in the sense of maximizing detection probability for the sweep rate assumed. The fine structure of the spread-spectrum signal is not specified, so rules for finding reasonable values of W and T require the computation of coincidence probabilities, swept signal time-bandwidth areas, and energy-detection probabilities for a large number of combinations of values of W, T, and signal strength. The difficulty in determining optimum values justifies seeking relatively simple rules for selecting good values of these parameters. The choice of rules given here was influenced somewhat by the fact that computations are very much simpler within certain ranges of parameter values.

Consider first the bandwidth W given in terms of r by (3.57). When the signal strength is low, a value of r less than $1/2$ would lead to unnecessarily small energy-detection probabilities. Similarly, if r is much larger than 1 and the amount of signal energy is small, the unnecessarily large amount of noise admitted would lead to small energy-detection probabilities. For larger SNR, one must consider the trade-off that occurs because the coincidence probability increases with W. For the weak-signal cases considered here, the best value of r will probably be between 0.7 and 1.0. Selecting several values of r in or near this region, one can compute the detection probability (as described later) for each of the several resulting values of W. The maximum of these detection probabilities will serve as an estimate of the interceptor's detection probability.

An integration time substantially less than the signal duration at the energy-detector input is suboptimum for weak signals. The formulas here are valid for integration times that exceed the signal duration at the energy-detector input, but matching the two generally results in higher detection probabilities for a fixed FAR.

Again inspecting Figures 2.21 and 2.24, we see that the integration time should be approximately

$$T = \min\left[T_2, \frac{W + W_2}{s}\right] \tag{3.61}$$

for the noncompression case, and

$$T_C = \min\left[\frac{W_2}{s}, T_2 + \frac{W}{s}\right] \tag{3.62}$$

for the compressive case. (The subscript C in (3.62) distinguishes between the two cases.) The minimum acceptable integration time is approximately $T_C = 1/W$. This is not pertinent for spread-spectrum signals, but for calculations involving a narrow-band signal with RF bandwidth less than s/W, let $T_C = 1/W$.

The signal energy actually accepted by the intercept receiver is E_I, and E_2 is the signal energy that is received if the sweep path completely covers the pulse. We let $\overline{E'}$ denote the intercepted energy averaged over all possible sweep paths under the condition $0 < E_I < E_2/2$, and $\overline{E''}$ denotes the average under the condition $E_I \geq E_2/2$. To estimate the ratio $\overline{E'}/E_2$ (or $\overline{E''}/E_2$) we assume that E_I/E_2 is equal to the ratio of the signal time-bandwidth area actually swept to the total time-bandwidth area T_2W_2 of the spread-spectrum pulse. The formulas for these averages are

$$\begin{aligned}\frac{\overline{E'}}{E_2} &= \frac{(1/4) + (1/12)q^2}{1 + q}, \quad q \leq 1 \\ &= \frac{(1/4) + (1/12)q^{-2}}{1 + q^{-1}}, \quad q > 1\end{aligned} \tag{3.63}$$

and

$$\frac{\overline{E''}}{E_2} = 1 - \frac{\overline{E'}}{rE_2} \tag{3.64}$$

where

$$q = \frac{W_2}{sT_2} \tag{3.65}$$

and (from (3.45) and the restriction on W):

$$r = \frac{W}{W_2 + sT_2} \geq \frac{1}{2} \tag{3.66}$$

The probability p_s of coinciding with and energy detecting any particular pulse is approximately

$$p_s \approx p' Q_D(\overline{E'}) + p'' Q_D(\overline{E''}) \tag{3.67}$$

with the energy-detection probability Q_D also a function of the integration time. (Q_D is discussed in Section 3.2.) Formulas for near-optimum values of the integration time are given by (3.61) for the conventional receiver and by (3.62) for the compressive receiver.

For the S-ED-T or S/PC-ED-T systems, the probability of detecting at least one of the b_2 pulses is

$$P_d = 1 - (1 - p_s)^{b_2} \tag{3.68}$$

If a false alarm occurs during the period of length T_1 when the transmission is arriving, or within a period $T_1/2$ before or after the transmission, it is likely that the pulses can be detected in a recognition stage. The probability of detecting the transmission is approximated by

$$P_d = 1 - (1 - p_s)^{b_2}(1 - 2T_1 \times \text{FAR}) \tag{3.69}$$

where p_s is given by (3.67), and the FAR is discussed below. Independence among pulse detections is assumed in (3.68) and (3.69).

The FAR for the S-ED-T system is FAR $= Q_F/T$ for I+D energy detection and FAR $\approx 2Q_F/T$ for continuous integration. (Q_F is discussed in Section 3.2.) The formula for FAR is similar in the compressive case, with T_C replacing T. The difference is that there is a period of length W/s during which no decision is made because swept frequencies outside the surveillance band are conveniently compressed into a single block. This can be observed in Figure 2.24. In the noncompressive case the energy from outside the surveillance band cannot be separated from the energy inside, as can be seen in Figure 2.21. The effect of the reduced decision rate in the compressive case is minor when Q_F is small. The FAR for I+D integration for the S/PC-ED-T system is FAR $= (W_s/W_0)(Q_F/T_C)$, and is approximately doubled for continuous integration.

Section 3.6.2 will discuss the use of formulas given earlier for calculating the FAR and the detection probability P_d for the S-ED-BMW and S/PC-ED-BMW systems. The discussions below apply to these systems and to the S-ED-T and S/PC-ED-T systems.

The remarks in Tables 2.9 through 2.12 on relationships of parameters assume that the interceptor using frequency scanning chooses the type of system and the parameter values that provide the best intercept capability. One consideration is whether to assume a compressive or noncompressive receiver. Some guidelines for making this choice are given below.

Consider a system having specified values of the sweep slope s, the bandwidth W, and the surveillance band W_s. The total noise energy integrated per decision is $N_{01}W \times$ (integration time). If the integration time is reduced and the same amount of signal energy is integrated, then the energy-detection probability Q_D is increased, even when the threshold of the energy detector is adjusted to maintain the same FAR. An exception to this increase in Q_D occurs only when the signal is so weak that Q_D is close to Q_F. The effect of varying the integration time while maintaining a fixed FAR can be observed in Graph Set II.

When the compressive system and noncompressive system are using the same sweep slope, they require approximately the same bandwidth for optimum detection of a weak spread-spectrum pulse of specified dimensions (optimum in the sense of maximizing the detection probability). Only a small amount of accuracy will be lost if we disregard the fact that the FAR is proportional to W_s/W_0 in the compressive case. If we also disregard any system losses attributable to the insertion loss of the dispersive device, then a comparison can be made between compressive systems and noncompressive systems on the basis of required integration time. Approximate formulas for the optimum integration time were given in (3.61) and (3.62). Based on the ratio T_C/T, compression is better if $s > W_2/T_2$ and noncompression is better if $s < W_2/T_2$. These results are summarized in Table 3.5, where it should be noted that $s < W_2/T_2$ in Case 1 and $s > W_2/T_2$ in Case 3.

The coincidence probability p'' in (3.59) increases very slowly with s for $s < W_2/T_2$ and rapidly with s for $s > W_2/T_2$. For weak signals, the situation where $E_I \geq E_2/2$ contributes most to the detection probability; that is, (3.67) is approximated by $p_s \approx p''Q_D(\overline{E''})$ when $Q_D(\overline{E''})$ is between about 0.1 and 0.9.

Recall that r is the ratio of the actual bandwidth W to the minimum bandwidth that would allow acceptance of the entire pulse. In the comparison in Table 3.5, r is assumed the same for both systems. For compressive receivers, the difficulty of implementing any particular r increases with s, because the time-bandwidth product of the dispersive device must be

$$W^2/s = r^2 W_2 T_2 \left(2 + \frac{W_2}{sT_2} + \frac{sT_2}{W_2} \right) \tag{3.70}$$

which increases almost linearly with s for $s \gg W_2/T_2$. Suppose that the interceptor is unable to implement a time-bandwidth product greater than C. By solving

$$C = W^2/s \tag{3.71}$$

for s and using (3.57), we have the result that the maximum sweep slope for a given value of r satisfies the equation:

$$s = \frac{r^2}{C}(W_2 + T_2 s)^2 \tag{3.72}$$

Table 3.5 A Comparison of Compressive and Noncompressive Systems on the Basis of the Ratio of Required Integration Times

	T	T_C	$\dfrac{T_C}{T}$	*Comparison*
Case 1: $sT_2 < W + W_2$, $W_2 > W + sT_2$	T_2	$T_2 + \dfrac{W}{s}$	$1 + \dfrac{W}{sT_2}$	Noncompression better.
Case 2: $sT_2 < W + W_2$, $W_2 < W + sT_2$	T_2	$\dfrac{W_2}{s}$	$\dfrac{W_2}{sT_2}$	Noncompression better if $s \ll W_2/T_2$; compression better if $s > W_2/T_2$.
Case 3: $sT_2 > W + W_2$	$\dfrac{W + W_2}{s}$	$\dfrac{W_2}{s}$ if $W \geq \dfrac{s}{W_2}$	$\dfrac{1}{(W/W_2) + 1}$	Compression better.

Solving this quadratic equation gives the maximum value of s:

$$s = \frac{W^2}{T_2^2}\left[\frac{C}{2r^2T_2W_2}\left(1 \pm \sqrt{1 - \frac{4r^2T_2W_2}{C}}\right) - 1\right] \tag{3.73}$$

The sign of the radical in (3.73) is negative for $s > W_2/T_2$ (the case where compression is better) and positive for $s < W_2/T_2$ (the case where noncompression is better). Note the requirement:

$$C \geq 4r^2T_2W_2 \tag{3.74}$$

For $r = 1/2$ this becomes $C \geq T_2W_2$, and for $r = 1$, $C \geq 4T_2W_2$. Letting $C = 4r^2T_2W_2$ gives $s = W_2/T_2$, the point at which the compressive and noncompressive systems perform nearly identically.

We have seen that if the communicator chooses $T_2W_2 \geq C$, where C is the maximum time-bandwidth product that the interceptor can implement, the compressive receiver cannot sweep over at least half the pulse's energy by using a sweep slope greater than W_2/T_2. In strong-signal cases, the interceptor can enjoy a high sweep rate and successful detection with a compressive receiver, even if only a small portion of the signal is swept. In the weak-signal case, he generally would perform better by using a conventional scanning system with $1/2 \leq r \leq 1$ and $s < W_2/T_2$.

Suppose the intercept threat is limited to a single compressive receiver with a fixed sweep slope s determined by the delay-frequency characteristics of the dispersive filter (Section 2.4). Also, suppose that the communicator has chosen a time-bandwidth product T_2W_2 less than the maximum time-bandwidth product $C = W^2/s$ of the interceptor's compression system. Therefore, the interceptor is able to implement a bandwidth W great enough to receive most of the pulse. If the communicator knows this value of s and he can efficiently transmit and receive pulses of dimension $T_2 = \sqrt{T_2W_2/s}$ and $W_2 = \sqrt{T_2W_2s}$, with T_2W_2 fixed, the communicator ensures that the interceptor will gain no advantage from pulse compression. The best that the frequency-scanning interceptor can do then is to find the value of s that maximizes his probability of coinciding with and detecting a pulse when not using compression. If the communicator cannot efficiently implement a pulse bandwidth wide enough that $W_2/T_2 = s$ (still assuming T_2W_2 has been selected), he should use the largest ratio W_2/T_2 practicable.

If frequency-scanning receivers with the option of pulse compression are the major signal interception threat, the conclusion above is suitable. Another consideration, however, is that the number of filters required in a filter-bank detection system decreases as the ratio W_2/T_2 is increased (with T_2W_2 fixed), because the bandwidths of the pulse and filter should be matched for best detection results.

3.6.2 Frequency Scanning and Binary Moving-Window Detection

Binary moving-window integration can be significantly better than simple thresholding when the number of coincidences of the sweep with pulses is large. If the pulses are time-hopped but not frequency-hopped, the binary moving-window systems are not applicable, unless a large number of scans occur during the transmissions.

The formulas given in Section 3.5 for the F-ED-BMW system apply to the S-ED-BMW and S/PC-ED-BMW systems (Figures 2.20 and 2.23) when p_s [given by (3.67)] is substituted for p_1 in (3.44) and when the window length is $M = T_1/T$ for the S-ED-BMW system [T given by (3.61)] and is $M = T_1/T_C$ for the S/PC-ED-BMW system [T_C given by (3.62)]. The same integration time (T or T_C) is used in the calculation of Q_D. The FAR is P_T/T for the S-ED-BMW system and is P_TW_s/W_0T_C for the S/PC-ED-BMW system.

3.6.3 Frequency Scanning over Elements

When the individual data symbols of the transmission consist of frequency-hopped elements, substitute T_3 for T_2, W_3 for W_2, and b_2b_3 for b_2 in the equations for binary moving-window detection. Alternatively, treat the data symbols as PN pulses. The equations in Section 3.6.1 assumed spread-spectrum pulses with signal energy distributed uniformly over an area T_2W_2. The equations are only approximations for pulses having unity *TW* product. The arguments for choosing between integration over a transmission and detection of individual data symbols also apply to the detection of hopped elements within a data symbol.

3.6.4 Multiple Observations of PN Transmissions

When the center frequency of the transmission is unknown to the interceptor, he can sweep across the band of possible transmission frequencies with a compressive receiver by using a sweep slope s such that his sweeping coincides or intersects with the transmission several times. Each time his system fully coincides with the transmission, it receives signal energy:

$$E_I = E_1 W / sT_1 \tag{3.75}$$

(See Figure 2.24, but assume that the transmission is a relatively long pseudonoise signal.) The scanning system will coincide with the transmission an average of T_1/T_s times, where $T_s = W_0/s$. The probability of detecting the transmission is

$$P_d = 1 - [1 - Q_D(E_I)]^{T_1 T_s} \tag{3.76}$$

3.7 SCANNING IN AZIMUTH

An interceptor can achieve a much greater antenna gain in the direction of the signal by using a highly directive antenna instead of an omnidirectional one. A narrow-beam system also provides location data and enables the detection of a weak signal when a stronger source is operating at another azimuth. To cover a wide angular surveillance area, however, the interceptor must either employ a large number of detection systems, each with its narrow-beam antenna pointed in a different direction, or he must time-share coverage by using one detection system and scanning its narrow beam in azimuth. We now consider the latter case.

Assume that the interceptor only receives energy through his main beam (of beamwidth θ_d), and he receives it uniformly in azimuth over the beam. He scans in azimuth over an angular sector of total angle θ_s either by mechanically rotating an antenna at a constant speed ($\theta_d \ll \theta_s = 360°$), or by using an array antenna to form successive beams, no two significantly overlapping in azimuth ($\theta_d \ll \theta_s \leq 360°$). For exactness, the time-direction coincidence formulas given in this section require that the time-order of beam positions be the same for every scan.

Figure 3.37 illustrates the antenna-beam sweep path for a mechanically rotating antenna. If the antenna rotates but a sector is *not* under surveillance, the false-alarm rate is reduced proportionally while the detection probability is computed as though $\theta_s = 360°$. The formulas are only approximately valid for mechanical sector scanning in alternate directions. Figure 3.38 illustrates the antenna-beam "sweep path" for an array.

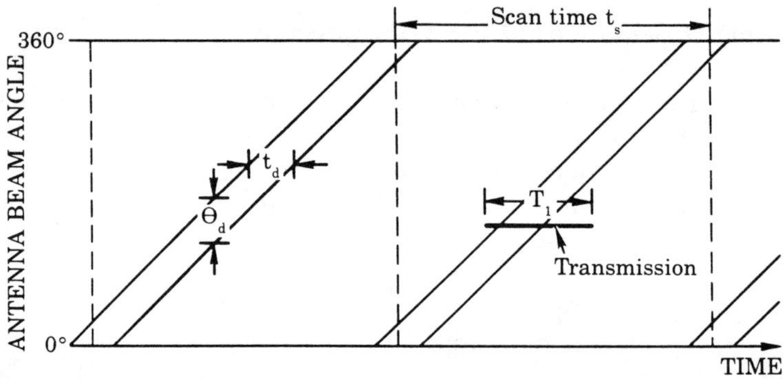

Figure 3.37 Antenna-beam sweep path for a mechanically rotating antenna having beamwidth θ_d.

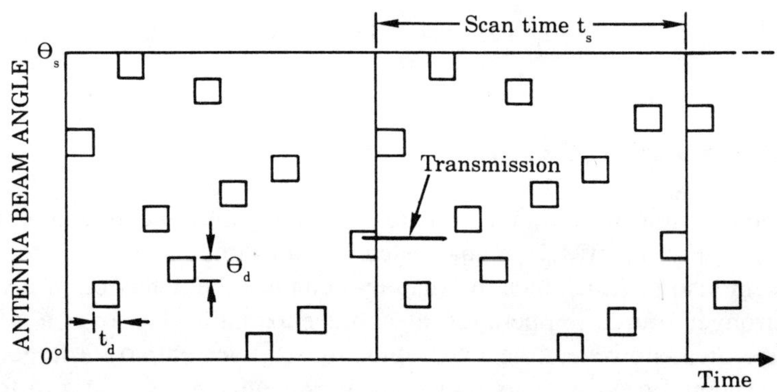

Figure 3.38 Coverage by an array.

The parameters associated with the antenna-beam sweep path are:

θ_d = interceptor's antenna beamwidth in azimuth;
θ_s = size of angular sector covered by interceptor each scan;
t_d = duration of interceptor's look in each direction or position (per scan);
t_s = duration of interceptor's scan.

From Figures 3.37 and 3.38 we see that

$$\frac{t_d}{t_s} = \frac{\theta_d}{\theta_s}$$

because an angular section having angle θ_s is scanned in time t_s. The parameters associated with the transmission are:

T_1 = transmission duration;
E_1 = total received energy from transmission during time T_1 if intercept antenna beam were to remain pointed at transmitter.

3.7.1 Signal Duration Less than Time between Looks

When $T_1 < t_s - t_d$, the interceptor can receive energy from the transmission during at most one scan. The maximum energy that he can intercept is

$$E_{\max} = \min\left[E_1, \frac{E_1 t_d}{T_1}\right] \tag{3.77}$$

The actual amount of energy that the interceptor will receive from the transmission is a random variable, which we will denote by E_I.

A rough approximation to the expected value $\overline{P_d}$ of the interceptor's detection probability is

$$\overline{P_d} \approx P_d(0.25E_{\max})Pr\{0 < E_I < 0.5E_{\max}\} + P_d(0.75E_{\max})Pr\{0.5E_{\max} \le E_I < E_{\max}\} + P_d(E_{\max})Pr\{E_I = E_{\max}\} \tag{3.78}$$

where P_d is the detection probability expressed as a function of the intercepted energy. The first term of this approximation usually is negligible if $P_d(E_{\max})$ is less than about 0.9. The formulas for time-direction coincidence probabilities are given in the list below. The detection probability P_d is calculated according to instructions (earlier in this chapter) corresponding to the particular detection system assumed. For the single-filter energy-detection system F-ED-T using continuous integration with integration time $T = \min[T_1, t_d]$, the detection probability P_d is simply the energy-detection probability Q_D and the false-alarm rate is

$$\text{FAR} \approx \frac{2Q_F}{T}$$

Time-Direction Coincidence Probabilities for $T_1 < t_s - t_d$

$$Pr\{E_I > 0\} = \frac{t_d + T_1}{t_s} \tag{3.79}$$

$$Pr\{E_I = 0\} = \frac{t_s - t_d - T_1}{t_s} \tag{3.80}$$

$$Pr\{E_I \geq 0.5E_1\} = \begin{cases} \dfrac{t_d}{t_s} & \text{if} \quad T_1 \leq 2t_d \\ 0 & \text{if} \quad T_1 > 2t_d \end{cases} \tag{3.81}$$

$$Pr\{E_I \geq 0.5E_{\max}\} = \begin{cases} \dfrac{t_d}{t_s} & \text{if} \quad T_1 \leq t_d \quad [E_{\max} = E_1] \\ \dfrac{T_1}{t_s} & \text{if} \quad T_1 \geq t_d \quad \left[E_{\max} = \dfrac{E_1 t_d}{T_1}\right] \end{cases} \tag{3.82}$$

$$Pr\{0 < E_I < 0.5E_1\} = \begin{cases} \dfrac{T_1}{t_s} & \text{if} \quad T_1 \leq 2\,t_d \\ \dfrac{t_d + T_1}{t_s} & \text{if} \quad T_1 > 2t_d \end{cases} \tag{3.83}$$

$$Pr\{0 < E_I < 0.5E_{\max}\} = \begin{cases} \dfrac{T_1}{t_s} & \text{if} \quad T_1 \leq t_d \quad [E_{\max} = E_1] \\ \dfrac{t_d}{t_s} & \text{if} \quad T_1 \geq t_d \quad \left[E_{\max} = \dfrac{E_1 t_d}{T_1}\right] \end{cases} \tag{3.84}$$

$$Pr\{E_I = E_1\} = \begin{cases} \dfrac{t_d - T_1}{t_s} & \text{if} \quad T_1 \leq t_d \\ 0 & \text{if} \quad T_1 \geq t_d \end{cases} \tag{3.85}$$

$$Pr\{E_I = E_{\max}\} = \begin{cases} \dfrac{t_d - T_1}{t_s} & \text{if} \quad T_1 \leq t_d \quad [E_{\max} = E_1] \\ \dfrac{T_1 - t_d}{t_s} & \text{if} \quad T_1 \geq t_d \quad \left[E_{\max} = \dfrac{E_1 t_d}{T_1}\right] \end{cases} \tag{3.86}$$

$$Pr\{0.5E_1 \le E_I \le E_1\} = \begin{cases} \dfrac{T_1}{t_s} & \text{if} \quad T_1 \le t_d \\ \dfrac{t_d}{t_s} & \text{if} \quad t_d \le T_1 \le 2t_d \\ 0 & \text{if} \quad T_1 > t_d \end{cases} \tag{3.87}$$

$$Pr\{0.5E_{\max} \le E_I < E_{\max}\} = \begin{cases} \dfrac{T_1}{t_s} & \text{if} \quad T_1 \le t_d \quad [E_{\max} = E_1] \\ \dfrac{t_d}{t_s} & \text{if} \quad T_1 \ge t_d \quad \left[E_{\max} = \dfrac{E_1 t_d}{T_1}\right] \end{cases} \tag{3.88}$$

3.7.2. Signal Duration Greater than Time between Looks

When $T_1 > t_s - t_d$, the interceptor is certain to receive some of the signal energy. The maximum amount of energy received from the signal during one scan is

$$E_{\max} = \frac{E_1 t_d}{T_1} \tag{3.89}$$

When the signal duration is less than the scan period ($t_s - t_d < T_1 < t_s$), the probability of receiving the maximum signal energy is

$$Pr(E_I = E_{\max}) = \frac{T_1 - t_d}{t_s} \tag{3.90}$$

The detection probability can be approximated by

$$P_d \ge Q(E_{\max}) \frac{T_1 - t_d}{t_s} \tag{3.91}$$

When the signal duration exceeds the scan period ($T_1 > t_s$), the number of time-direction coincidences will be an integer close to

$$b_c = \frac{T_1}{t_s} \tag{3.92}$$

although on the first or last of these coincidences the interceptor might intercept less than the maximum amount of energy. If the interceptor uses a single-filter energy-detection system (F–ED-T) with integration time $T_I = t_d$ and continuous integration, his probability of intercepting the transmission is

$$P_d \approx 1 - [1 - Q_D(E_{\text{max}})]^{b_c} \tag{3.93}$$

and his false-alarm rate is

$$\text{FAR} \approx \frac{2Q_F}{t_d}$$

If b_c is large, the interceptor can improve his detection probability by using scan-to-scan interception or by using appropriate logic in his alarm circuit.

REFERENCES

1. Skolnik, M. I., *Introduction to Radar Systems,* 2nd Ed., New York: McGraw-Hill, 1980.
2. Schwarz, M., *Information, Transmission, Modulation, and Noise,* 2nd Ed., New York: McGraw-Hill, 1970.
3. Urkowitz, H., "Energy Detection of Unknown Deterministic Signals," *Proc. IEEE,* Vol. 55, No. 4, April 1967, pp. 523–531.
4. Torrieri, D. J., *Principles of Secure Communication Systems,* Norwood, MA: Artech House, 1985.
5. Dillard, G. M., "Recursive Computation of the Generalized *Q* Function," *IEEE Trans. on AES,* Vol. AES-9, No. 4, July 1973, pp. 614–615.
6. Helstrom, C. W., "Approximate Evaluation of Detection Probabilities in Radar and Optical Communications," *IEEE Trans. on AES,* Vol. AES-14, No. 4, July 1978, pp. 630–640.
7. Abramowitz, M., and I. A. Stegun, eds., *Handbook of Mathematical Functions,* National Bureau of Standards, AMS-55, Seventh Printing, May 1968.
8. Pachares, J., "A Table of Bias Levels Useful in Radar Detection Problems," *IRE Trans. on Inf. Theory,* Vol. IT-4, March 1958, pp. 38–45.
9. Robertson, G. H., "Computation of the Noncentral Chi-Square Distribution," *Bell System Tech. Journal,* Vol. 48, January 1969, pp. 201–207.
10. Parl, S., "A New Method of Calculating the Generalized *Q* Function," *IEEE Trans. on Inf. Theory,* Vol. IT-26, No. 1, January 1980.
11. DiFranco, J. V., and W. L. Rubin, *Radar Detection,* Norwood, MA: Artech House, 1980.
12. Barton, D. K., "Simple Procedures for Radar Detection Calculations," *IEEE Trans. on AES,* Vol. AES-5, September 1969, pp. 837–846.
13. Cann, A. J., "Simple Radar Detection Calculations," *IEEE Trans. on AES,* Vol. AES-8, January 1972, pp. 73–74.
14. Urkowitz, H., "Closed-Form Expressions for Noncoherent Radar Integration and Collapsing Loss," *IEEE Trans. on AES,* Vol. AES-9, September 1973, pp. 781–783.
15. Dillard, R. A., "Detectability of Spread-Spectrum Signals," *IEEE Trans on AES,* Vol. AES-15, No. 4, July 1979, pp. 526–537.
16. Tiuri, M. E., "Radio Astronomy Receivers," *IEEE Trans. on A&P,* Vol. AP-12, December 1964, pp. 930–938.
17. Ohlson, J. E., "Efficiency of Radiometers using Digital Integration," *Radio Science,* Vol. 6, No. 3, May 1970, pp. 341–345.
18. Hansen, V. G., "A Sequential Logic for Improving Signal Detectability in Frequency-Agile Search Radars," *IEEE Trans. on AES,* Vol. AES-4, September 1968, pp. 763–773.
19. Dillard, G. M., "Binary Detection of Randomly Occurring Signals," *Naval Electronics Laboratory Center,* Technical Report TR-1741, 30 November 1970 (unlimited distribution).
20. Wiley, R. G., *Electronic Intelligence: The Interception of Radar Signals,* Norwood, MA: Artech House, 1985.

Chapter 4
Effect of Signal Structure on Detectability

In this chapter, we compare the appropriate detection systems for each of the major types of spread-spectrum signal structures and give graphs and instructions for selecting optimum parameters. Later, we reach some general conclusions about the relative detectability of the various structures.

The signal parameters used in this chapter are

W_1 = transmission bandwidth;
T_1 = transmission duration;
W_2 = pulse bandwidth;
T_2 = pulse duration;
T_{p2} = duration of hop interval;
b_2 = number of pulses (data symbols) in time T_1;
E_1 = total received signal energy in time T_1;
E_2 = received signal energy per pulse ($= E_1/b_2$);
N_{01} = one-sided noise power spectral density.

The interceptor parameters are

W = filter bandwidth;
T = energy integration time;
N = number of filters;
P_d = probability of detecting transmission;
FAR = false-alarm rate (alarms per second).

The hopping pattern of pulses over the time-bandwidth area of a transmission affects detectability very little. FH and TH transmissions are nearly equally detectable if they have the same time-bandwidth areas, number of pulses, and energy. It is convenient, therefore, to compute detection curves as a function of what we will call the *density*. The ratio of the time-bandwidth area $b_2T_2W_2$ occupied by the b_2 pulses (data symbols) to the area T_1W_1 of the transmission is its density, denoted by D:

$$D = b_2T_2W_2/T_1W_1 \tag{4.1}$$

The density becomes $D = b_2T_2/T_1 = T_2/T_{p2}$ if the hopping is over time but not frequency. If the hopping is over frequency but not time, the density is $D = W_2/W_1$. In the latter case, if the transmission band is covered by N nonoverlapping filters matched to the pulse bandwidth, we have $D = 1/N$.

We will give curves of detectability below as a function of D. When D is sufficiently large, energy integration over the transmission (F-ED-T system with $T = T_1$ and $W = W_1$) will require the least signal energy E_1 for specified interceptor performance. When the pulses are sparsely distributed, detection methods that are based on detection of individual pulses are generally better than integration over the transmission. We will see later that the product b_2D is often a good indicator of when the two detection methods perform equally.

To compare pulse-detection systems that employ a binary moving-window detector (BMWD) with pulse-detection systems that alarm on a single threshold crossing, we introduce the notion of a *processing gain*. The processing gain G_{MW} is the difference (in dB) between the signal energy $E_{1,\mathrm{Thr}}$ required by a simple pulse-detection system (the F-ED-T or FB-ED-T system with $T = T_2$) and the energy $E_{1,\mathrm{BMW}}$ required by the BMWD system appropriate against that signal (the F-ED-BMW, F-ED-OR/BMW, FB-ED-BMW, or FB-ED-OR/BMW system). The processing gain is given by the following expression, where the same values of T, W, N, P_d, and FAR are assumed for both the simple system and the BMWD system:

$$G_{\mathrm{MW}} = 10\log_{10}(E_{1,\mathrm{Thr}}/N_{01}) - 10\log_{10}(E_{1,\mathrm{BMW}}/N_{01}) \tag{4.2}$$

Alternatively, the pulse energy E_2 can be used. The signal energy required by a system employing a BMWD can be estimated by computing the energy-to-noise ratio the simple thresholding system requires and adjusting that value with an estimate of the processing gain G_{MW} applicable for the specified signal parameters and detection performance. The gain G_{MW} should also serve as a measure of the effectiveness of other nearly optimal alarm circuits.

4.1 TH AND TH/PN TRANSMISSIONS

In this section, we compare the effectiveness of several systems against time-hopped (TH) transmissions. These may be pure TH transmissions or TH hybrid transmission (see Section 1.2), in which case the time-hopped pulse has a spread-spectrum microstructure (e.g., pseudonoise, PN). Figure 4.1 shows an example of a TH hybrid such as TH/PN. In Figure 4.1, the bandwidth of the transmission is that of the pulse. However, the graphs and other results are valid for $W_2 \leq W_1 \leq W$.

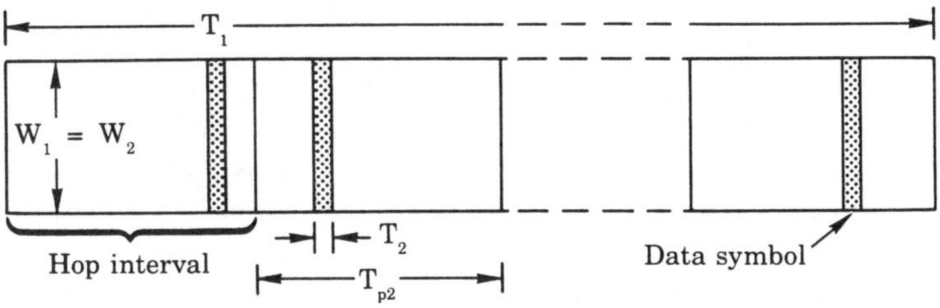

Figure 4.1 A time-frequency representation of time-hopped spread-spectrum pulses.

4.1.1 Detection System Comparisons

An interceptor with knowledge of the signal structure and parameters, except the pseudorandomly varied ones, can often increase the detection probability of a pulse-detection system by employing appropriate logic in the alarm circuit, as discussed in Chapter 2. One of the simpler alarm circuits effective against TH transmissions is a BMWD. In particular, we consider the F-ED-BMW system (see Figure 2.12 and Table 2.4) and, for computational reasons, the synchronized F-ED-OR/BMW system (described at the end of Section 2.2.3). Calculations of detection performance based on a BMWD should give a reasonable measure of the effectiveness of any alarm circuit of moderate complexity that was appropriately designed to operate against TH transmissions. Equations for BMWD computations were given in Section 3.5.

Simple energy integration over the entire transmission is sometimes better than any method based on individual pulse decisions, so vulnerability to both kinds should be considered. Figures 4.2 and 4.3 compare the performance of several detection systems for FAR $= W_1 \times 10^{-11}$ and FAR $= W_1 \times 10^{-8}$, respectively, where $T_1W_1 = 10^5$. In both figures, the top four curves give the vulnerability of a TH or TH/PN transmission having 100 pulses to (1) energy integration over the entire transmission (F-ED-T with $T = T_1$), (2) energy integration over pulses (F-ED-T with $T = T_2$), (3) integration over pulses followed by a BMWD (F-ED-BMW), and (4) integration over pulses followed by an OR-gate aligned in time with the hop interval and a BMWD (F-ED-OR/BMW). The fifth curve in both graphs is for integration of 100 periodically spaced pulses by collapsing in time (F-ED-C/T). The F-ED-OR/BMWD curve is unrealistic in that the interceptor will be aligned in time only by chance, but it is easier to compute than the F-ED-BMW curve (because the window length M is much smaller) and serves as an approximation. In all cases, the ordinate gives the value of E_1/N_{01} required by the interceptor's system under the conditions that $W = W_1$ and $P_d = 0.5$. In most cases, the bandwidth of the transmission will be the bandwidth of the pulse (i.e., $W_1 = W_2$). For $W_1 = W_2$, the density is $D = T_2/T_{p2}$. The density of the transmissions in Figures 4.2 and 4.3 is $D = T_2W_2 \times 10^{-3}$. For the F-ED-OR/BMW system, the window length is $M = b_2$ and the BMWD false-alarm probability P_T is 10^{-8} for Figure 4.2 and is 10^{-5} for Figure 4.3. The computation of p_1, Q_D, and Q_F used (3.46), (3.47), and (3.54), respectively. For the F-ED-BMW system, the window length is $M = 10^5/T_2W_1$ and $p_0 = Q_F$. The false-alarm probability P_T is $T_2W_1 \times 10^{-11}$ for Figure 4.2 and is $T_2W_1 \times 10^{-8}$ for Figure 4.3. The detection probabilities p_1 and Q_D were approximated by using (3.46) and (3.47), but spot checks were made by using the exact expression (3.48) with $p_1 = Q_D$. The exact results were slightly higher than the curve shown, but not usually noticeably so.

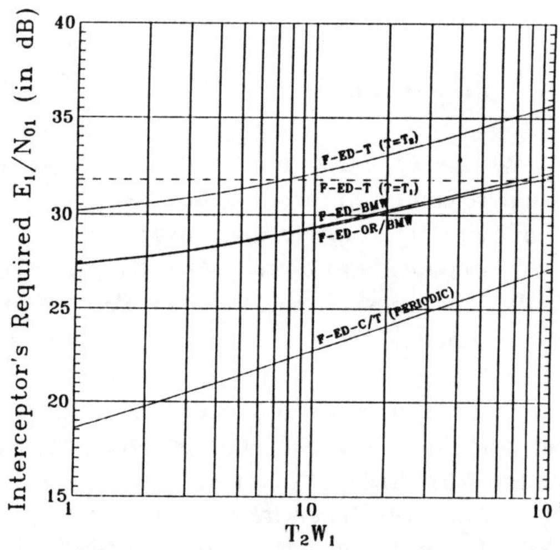

Figure 4.2 A comparison of pulse-detection systems for TH/PN transmissions; $b_2 = 100$, $T_1W_1 = 10^5$, $P_d = 0.5$, FAR $= W_1 \times 10^{-11}$.

Computations for systems with moderately complex alarm circuits such as the BMWD are much more difficult than computations for the F-ED-T system. If rough estimates of detectability or of optimum parameters are acceptable, one can calculate the energy (in dB) required by the F-ED-T system (with $T = T_2$), subtract an estimate of the processing gain G_{MW}, and use the resulting number or curve for the F-ED-BMW system or other fairly dedicated pulse-detection system. With this in mind, curves of G_{MW} are given in Figure 4.4 for a variety of parameter combinations.

The processing gain G_{MW} in Figure 4.4 is the dB difference between the signal energy required by the F-ED-T system ($T = T_2$) and that required by the F-ED-

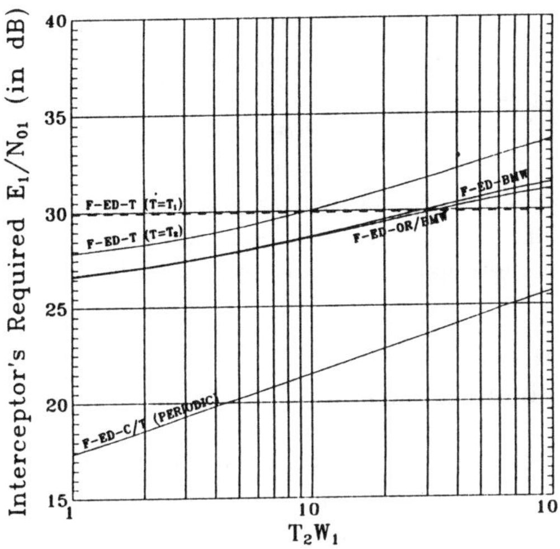

Figure 4.3 A comparison of pulse-detection systems for TH/PN transmissions; $b_2 = 100$, $T_1W_1 = 10^5$, $P_d = 0.5$, FAR $= W_1 \times 10^{-8}$.

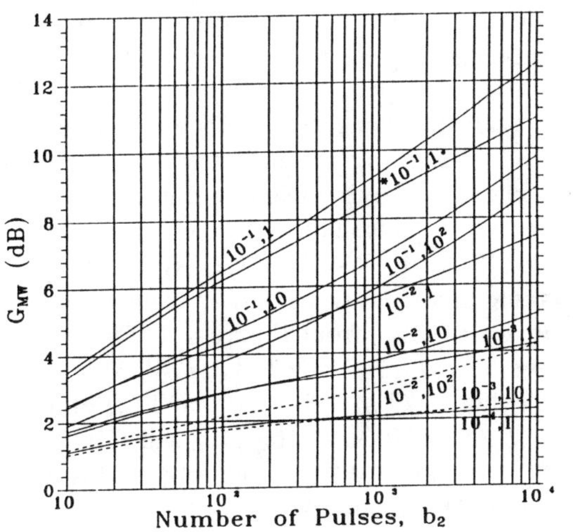

Figure 4.4 Processing gain G_{MW} of the F-ED-BMW system (marked by asterisk) and the F-ED-OR/BMW system for TH and TH/PN transmissions; $T = T_2$, $W = W_1 \geq W_2$, $P_d = 0.5$, FAR $= W_1 \times 10^{-11}$, $T_1W_1 = b_2(T_{p2}/T_2)(T_2W_1)$. Curve label is $[T_2/T_{p2}]$, $[T_2W_1]$, that is, label gives duty cycle and integration cell size.

BMW system (the curve noted by an asterisk) or the synchronized F-ED-OR/BMW system. Both the simple system and the BMWD system are assumed to have the same integration time, bandwidth, false-alarm rate, and detection probability. As pointed out in the discussion of Figures 4.2 and 4.3, the curves for the synchronized system approximate the performance of the F-ED-BMW system. Parameters for the F-ED-BMW curve are $M = b_2T_{p2}/T_2 = 10b_2$, $P_T = 10^{-11}$, and $p_0 = Q_F$. Those for the FB-ED-OR/BMW curves are $M = b_2$ and $P_T = (T_2W_1T_{p2}/T_2) \times 10^{-11}$. Equation (3.54) gives the relationship of p_0 and Q_F. Equations (3.46) and (3.47) were used to find p_1 and Q_D for both systems. Because these are approximations for the F-ED-BMW curve, several calculations with (3.44) and $p_1 = Q_D$ were made to check their accuracy.

4.1.2. Selection of Optimum Parameters

The important information to be found in graphs such as Figures 4.2 and 4.3 concerns optimization of signal parameters. First, it is useful to note in Figures 4.2 and 4.3 that increasing the pulse time-bandwidth product, by increasing T_2, beyond the point where the curve for energy integration over the transmission and the best curve for pulse detection cross does not improve resistance to detection by a dedicated interceptor. Therefore, choosing T_2W_1 ($W_1 = W_2$ fixed) at that point where the two curves cross is the best strategy for low detectability. Using a smaller value increases detectability, whereas using a larger value increases implementation complexity (a longer pseudorandom sequence segment is needed) without gain in detection resistance.

We reiterate that, against a dedicated interceptor, the detectability of TH/PN transmissions cannot be decreased by increasing the pulse time-bandwidth product beyond the point where integration over the entire transmission (F-ED-T system with $T = T_1$) becomes as effective as pulse detection combined with a good alarm circuit (modeled by the F-ED-BMW or F-ED-OR/BMW system with $T = T_2$). Figure 4.5 illustrates how to determine this crossover point by using a figure (Figure 3.29) from Graph Set III and estimates of G_{MW} from Figure 4.4. First note that, for $W_1 = W_2$ and $T_1W_1 = 4 \times 10^5$, we have $T_2/T_{p2} = b_2T_2W_2/T_1W_1 = T_2W_2/2000$. For $T_2W_2 = 10$, for example, we calculate $T_2/T_{p2} = 5 \times 10^{-3}$ and therefore interpolate for G_{MW} between the curves for $(T_2/T_{p2}, T_2W_2) = (10^{-2}, 10)$ and $(10^{-3}, 10)$ in Figure 4.4. Interpolation for $T_2W_2 = 1$, 10, and 100, respectively, yields approximate values of G_{MW} of 2.5, 2.7, and 3.5. The required value of E_1/N_{01} for IOT (dashed curve) intersects the adjusted curve at about $T_2W_2 = 80$, which is an estimate of the best pulse size.

In addition to the gain G_{MW}, other gains or losses peculiar to only one of the two detection systems should enter into the estimate of the dB adjustment. When using Graph Set III curves for TH/PN cases, it is useful to note that $T_{p2}/T_2 = T_1W/b_2T_2W$. One usually assumes $W = W_1$ in the time-hopping situations considered.

In the example shown in Figure 4.5, the best value of T_2 is determined for fixed T_1 and W_1, where usually $W_1 = W_2$. In a similar manner, one can fix T_2 and graphically determine for what value of T_1 (or, equivalently, of T_{p2}) the two types of detection are matched in performance. When using Graph Set III and estimates of gain for this situation, several iterations in estimating G_{MW} are needed, because it is a function of T_{p2}/T_2.

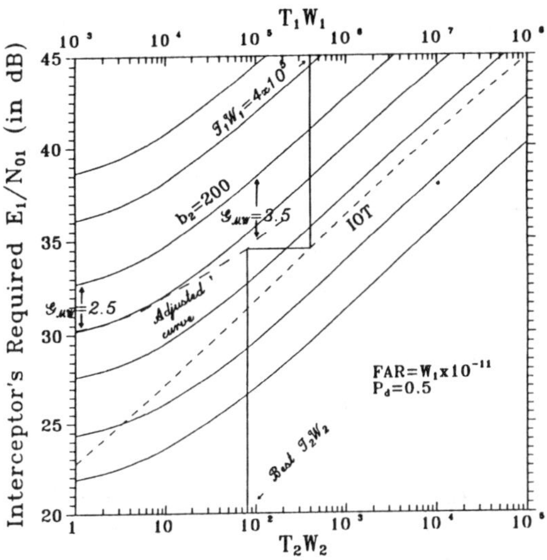

Figure 4.5 Example of how to estimate the performance of a BMWD system and compare it with integration over the transmission, by using curves from Graph Set III and estimates of G_{MW}. Crossover point indicates best pulse size for $T_1W_1 = 5 \times 10^5$.

An illustration of the effect of increasing T_1 for a pure FH transmission is given in Figure 4.6, adapted from Dillard ([1], Figure 8). The hop interval duration T_{p2} is T_1/b_2 and the density (or duty cycle) is $D = T_2/T_{p2} = b_2T_2/T_1 = b_2 \times 10^{-7}/T_1$. Curves for BMWD detection assume an F-ED-OR/BMW (synchronized) system having the optimum value of the threshold K_M. Curves for simple pulse detection (F-ED-T system with $T = T_2$) are essentially the same as for a BMWD system with $K_M = 1$. The point where the curve for integration over the transmission crosses the corresponding curve for the F-ED-OR/BMW system indicates the best value of T_1. The curves begin at values of T_1 such that the ratio T_2/T_{p2} is about 0.1. The curves for the F-ED-OR/BMW system use $M = b_2$, $P_T = (T_1/b_2) \times 10^{-4}$, and (3.46), (3.47), and (3.54).

The number of combinations of parameters for which calculations of G_{MW} were made is unfortunately small here, owing to the tediousness of computations. For TH/PN transmissions in general, G_{MW} tends to increase as the density $D = T_2/T_{p2}$, the pulse size T_2W_2, or the FAR decreases.

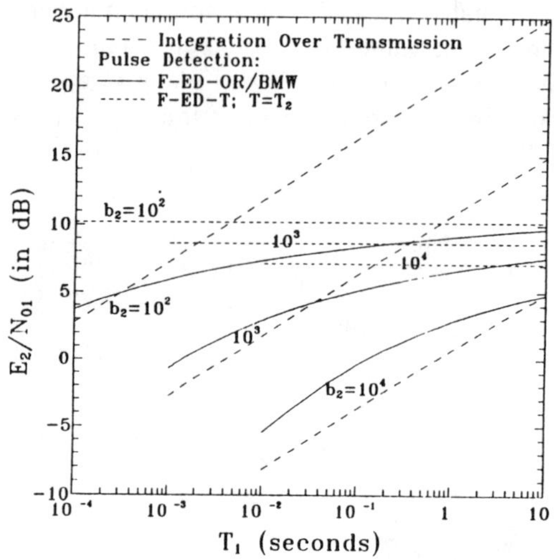

Figure 4.6 Effect of increasing the transmission time (by increasing the hop interval) of a pure TH transmission; $W = W_1 = W_2 = 10$ MHz, $T_2 = 0.1$ μs, FAR $= 10^{-4}$, $P_d = 0.5$. *Source:* After Dillard [1], Figure 8.

It is important to recognize that calculations for the F-ED-OR/BMW system (the OR operation over the hop interval) are identical to those for the FB-ED-OR/BMW system (the OR operation over frequency channels) when T_{p2}/T_2 is substituted for N in the FB-ED-OR/BMW detection formulas. Because of this relationship, the graphs of G_{MW} given in Section 4.2 and all of the results given in Section 4.4 are also valid for time hopping with perfect time alignment. In Section 4.2, substitute T_{p2}/T_2 for N, and note that the density becomes $D = T_2/T_{p2}$. Disregard N in bandwidth relationships in Section 4.2 and assume $W = W_1 \geq W_2$ for all systems being compared. When using results for the time-synchronized F-ED-OR/BMW system from Section 4.2 or Section 4.4, allow a slightly poorer performance for the F-ED-BMW system.

4.2 PURE FH TRANSMISSIONS

Recall that in a pure FH transmission, the data symbol is a simple pulse, and the pulses are contiguous in time (Section 1.2). The transmission duration is $T_1 = b_2T_2$ and the total energy is $E_1 = b_2E_2$. The appropriate detection systems are the F-ED-T ($T = T_1$), FB-ED-T ($T = T_2$), and FB-ED-OR/BMW systems. Figure 4.7 illustrates that as the number of possible frequencies increases, the performance of the filter-bank pulse-detection systems improves relative to energy integration over the transmission. However, the splitting losses increase the required signal energy for filter-bank systems. The BMWD false-alarm probability in Figure 4.7 is $P_T = 10^{-7}$. The false-alarm probability for pulse detection with the F-ED-T system is $Q_F = 10^{-7}/N$ and for IOT is $Q_F = b_2 \times 10^{-7}$.

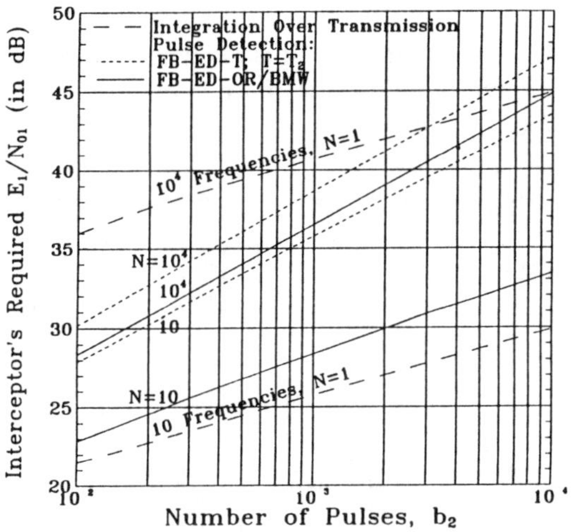

Figure 4.7 A comparison of detection methods for pure FH transmissions; N = number of filters, $N = W_1/W_2$ for pulse detection, $T_1W_1 = b_2 \times$ (number of frequencies), FAR $= W_2 \times 10^{-7} = W_1 \times 10^{-7}$/(number of frequencies), $P_d = 0.5$.

Figure 4.8 gives curves of the BMWD processing gain G_{MW}. To estimate the amount of signal energy required by the FB-ED-OR/BMW system for a specified detector performance, find the energy required by the FB-ED-T system (with $T = T_2$) by using the formulas in Table 3.4 or curves in a Chapter 3 graph set, and subtract an estimate of G_{MW}.

The results given in later sections for the $T_2W_2 = 1$ case also apply to pure FH. The density given there as $D = b_2T_2W_2/T_1W_1$ becomes $D = W_2/W_1$ because $T_1 = b_2T_2$. For $N = W_1/W_2$ nonoverlapping filters, the density is $D = 1/N$.

4.3 FH/PN TRANSMISSIONS

The detection systems F-ED-T, FB-ED-T, and FB-ED-OR/BMW are applicable to FH/PN transmissions. The results given in Section 4.4 also apply. The density is $D = W_2/W_1$ under the usual assumption that the pulses are contiguous in time, and no two occur at the same time. Optimization of the pulse size T_2W_2 in Section 4.4 is based on the premise that signal detectability cannot be decreased by increasing the T_2W_2 beyond the point where energy integration over the entire transmission (FB-

ED-T system with $W = W_1$ and $T = T_1$) becomes as good as integration over pulses, combined with a moderately complex alarm circuit, modeled by the FB-ED-OR/BMW system with $W = W_2$, $T = T_2$, and $N = W_1/W_2$. If the transmission size T_1W_1 is not sufficiently large for the number of pulses it contains, energy integration over the transmission will be the best detection method, even if $T_2W_2 = 1$ (the pure FH case treated in Section 4.2). Figure 4.5 illustrated the optimization procedure, and the discussion for that figure applies here if we substitute N for T_{p2}/T_2. Additional curves of G_{MW} are given in Section 4.4.

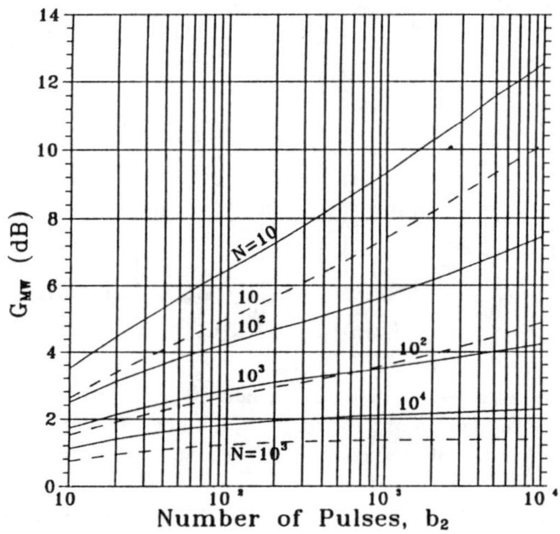

Figure 4.8 Processing gain for pure FH transmissions; N = number of filters = W_1/W_2, $T_1W_1 = b_2N$, $W = W_2$, $T = T_2$, $P_d = 0.5$. For solid curves FAR = $W_1 \times 10^{-11}$; for dashed curves FAR = $W_1 \times 10^{-8}$. *Source:* After Dillard [1], Figure 9.

4.4 FH/TH AND FH/TH/PN TRANSMISSIONS

The discussion in this section applies to the detection of transmissions consisting of pulses hopped in time and frequency, where the actual pulses or data symbols can be spread-spectrum signals. Although most of the discussion is only valid for FH or TH, the earlier sections dealing specifically with them ought to be read first, if those are the signals of interest.

Table 4.1 lists the detection systems for which we give results in this chapter. The results for the time-synchronized versions of the BMWD systems assume that a single pulse occurs in each hop interval. The results for the FB-ED-OR/BMW system assume that no two pulses occur at the same time. The FB-ED-T results are valid also for pulses occurring simultaneously at different frequencies.

Table 4.1 Detection Systems Applicable to FH/TH and FH/TH/PN Transmissions

Detection Technique	*Detection System*	*Reference and Comments*
Energy integration over entire transmission ($T = T_1$, $W = W_1$).	F-ED-T	Figure 2.9 Table 3.3 Section 3.4.2
Energy integration over individual pulses ($T = T_1$); simple thresholding.	FB-ED-T; $W = W_2$	Figure 2.14 Table 3.4
See Section 3.4.4.	F-ED-T; $W = W_1$ (if W_1/W_2 not large)	Figure 2.9 Table 3.3
Energy integration over individual pulses ($T = T_2$); BMWD integration of pulse decisions.	FB-ED-BMW; $W = W_2$ FB-ED-OR/BMW; $W = W_2$	 Figure 2.15
See Section 3.5.	Synchronized FB-ED-OR/BMW; $W = W_2$	Use $N^* = NT_{p2}/T_{p2}$ for N in computations.
	F-ED-BMW; $W = W_1$ (if W_1/W_2 not large)	Figure 2.12
	F-ED-OR/BMW; $W = W_1$ (if W_1/W_2 not large)	Use $N^* = T_{p2}/T_2$ for N in computations for FB-ED-OR/BMW.

Unless the density D is sufficiently small, energy integration over the entire transmission will require the least signal energy E_1 for specified interceptor performance. When the pulses are distributed sparsely, detection methods that are based on detection of individual pulses are generally better than integration over the transmission. The simplest type of pulse-detection system against FH/TH and FH/TH/PN transmissions is the FB-ED-T system. (If the band of frequency hopping is relatively narrow, the F-ED-T system can also be used effectively as a pulse-detection system.) The FB-ED-OR/BMW system represents a dedicated interceptor.

Recall that there are two ways to use the OR operation of the FB-ED-OR/BMW system when time hopping is involved. In the unlikely situation where the interceptor is able to synchronize with the hop interval, he can utilize the information that only one pulse occurs during each hop interval. Most of the results given in this chapter are for the synchronized system because the computations are much simpler (the window length M is decreased by the ratio T_2/T_{p2}). These results are at least approximately valid for various combinations of FH, TH, and PN signaling.

When estimating detectability for transmissions involving FH, recall that unless the interceptor knows precisely the possible frequencies of the transmission, he will suffer a frequency-splitting loss L_f, discussed in Section 3.3.3. (If he overlaps his filters by one-half to avoid that loss, his FAR is approximately doubled.)

4.5 FINAL COMPARISONS

Computations for detection systems using BMWD integration or any other alarm logic of moderate complexity are difficult compared to those for the FB-ED-T and F-ED-T systems. As we suggested at the beginning of this chapter, the signal energy required by a BMWD system can be estimated by computing the amount required by the corresponding simple thresholding system and subtracting the processing gain G_{MW}.

For FH/TH transmissions (pulse size $T_2W_2 = 1$), Figure 4.9 gives curves of G_{MW} as a function of the number of pulses. The curves are for a range of values of density D, and are also valid for pure TH and pure FH transmissions. For transmissions involving time hopping, the computations assume synchronization with the hop interval. For TH transmissions (with $W_1 = W_2$), the density is T_2/T_{p2}. For FH transmissions, the density is $D = W_2/W_1$. For FH/TH transmissions, the density is $D = b_2/T_1W_1$. When $N = W_1/W_2$, we have $D = T_2/T_{p2}N$.

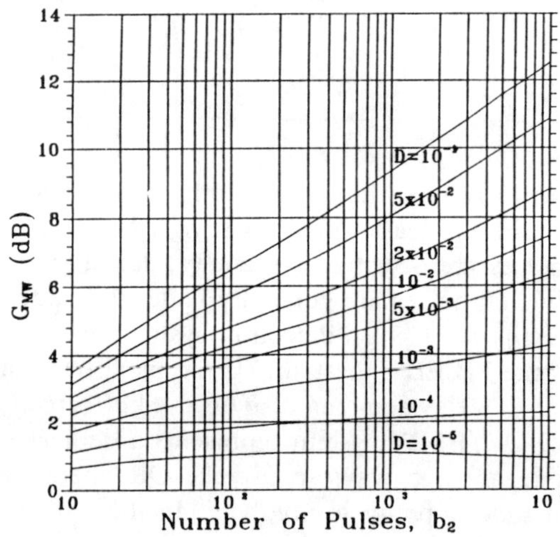

Figure 4.9 Processing gain for FH/TH transmissions; N = number of filters = W_1/W_2, $D = T_2/T_{p2}N$, $T_1W_1 = b_2/N$, $P_d = 0.5$, FAR $= W_1 \times 10^{-11}$

For FH and FH/TH transmissions, Figure 4.10 illustrates how an estimate of G_{MW} and the frequency-splitting loss L_f can be used with Figure 3.28 of Graph Set III to compare BMWD pulse-detection systems with integration over the transmission. The example is for 1000 pulses distributed over a time-bandwidth area T_1W_1

$= 5 \times 10^5$, resulting in density $D = 2 \times 10^{-3}$. From Figure 4.9, we find by interpolation that G_{MW} is about 4 dB. Figure 3.17 (with Figure 3.20) gives a value of about 0.9 dB for L_f. The adjustment to the value of required E_1/N_{01} is therefore 3.1 dB. The conclusion for this example is that the two detection methods perform essentially equally well, but integration over the transmission is the slightly better detection technique. The comparison technique illustrated in Figure 4.10 applies to pure TH transmissions by letting L_f be zero, in which case BMWD pulse detection is better.

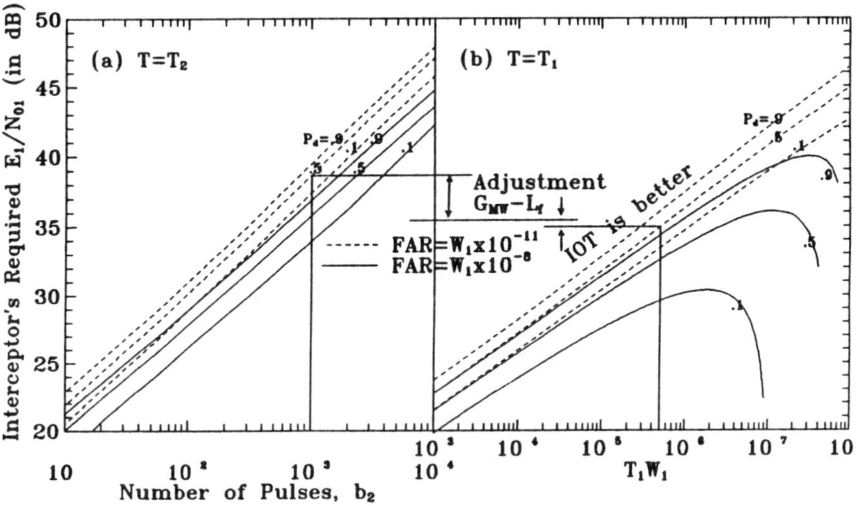

Figure 4.10 An example for FH, TH, and FH/TH transmissions of how to estimate the performance of a BMWD system and compare it with integration over the transmission, by using Figure 3.28 of Graph Set III and estimates of gain and loss.

Figure 4.11 gives curves of G_{MW} for combinations of D and T_2W_2, where $T_2W_2 \geq 10$. Values of G_{MW} for T_2W_2 between 1 and 10 can be found by interpolating between Figure 4.8 and Figure 4.9 for FAR $= W_1 \times 10^{-11}$. The curves in Figure 4.11 were computed for the time-synchronized FB-ED-OR/BMW system, and apply accurately to signal structures not involving time hopping. In general, G_{MW} increases as P_d increases or as FAR decreases, all other parameters remaining constant.

Figure 4.5 gave an example for TH/PN transmissions of how to find the best pulse size by using an estimate of G_{MW} with a curve from Graph Set III. This example and its result are applicable to FH/PN and FH/TH/PN transmissions if we replace the quantity T_2/T_{p2} with the density D. If the pulses are any larger than the value indicated, then the communicator has gone to unnecessary trouble in implementing the spread-spectrum pulses. Using a larger T_2W_2 would provide no additional protection from a dedicated interceptor because integration over the transmission would then be the better detection technique and its performance depends on T_1W_1 rather than T_2W_2. If the communicator uses a pulse size less than that which equalizes the two kinds of detection, he is not taking advantage of the fact that his transmission is spread over a large time-bandwidth area. The interceptor can use a good pulse-detection method and will not suffer as much degradation as he would by integrating noise energy $T_1W_1N_{01}$.

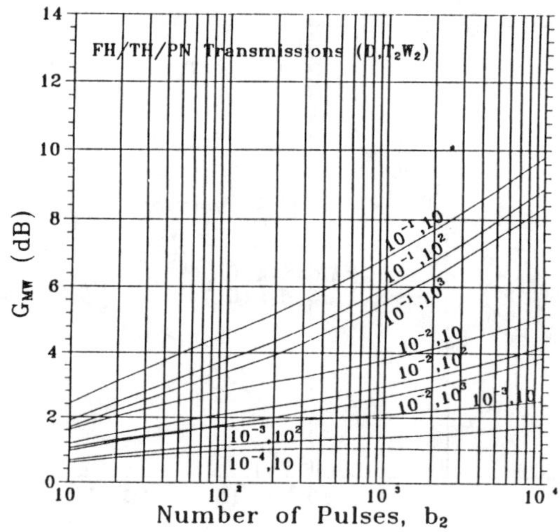

Figure 4.11 Processing gain for FH/TH/PN transmissions; N = number of filters = W_1/W_2, $W = W_2$, $T = T_2$, $D = T_2/T_{p2}\,N$, $P_d = 0.5$, FAR = $W_1 \times 10^{-11}$.

In the above discussion we have assumed that T_1W_1 is fixed (and W_1 should be fixed when using figures where FAR is proportional to W_1). It can also be that, for example, the signal parameters b_2, T_2W_2, and W_1 have been selected and the best value of T_1 (equivalently, the best value of the hop interval T_{p2}) is to be found. The discussion regarding the optimization of T_1 described for TH transmissions in Section 4.1.2 also applies to hybrid signals that involve time hopping.

Figures 4.12 and 4.13 give the difference between the signal energy E_1 required for integration over the transmission (IOT) and that required for integration over pulses with a BMWD. The comparisons in Figure 4.12 apply to FH, TH, and FH/TH transmissions ($T_2W_2 = 1$). The comparisons in Figure 4.13 apply to FH/PN, TH/PN, and FH/TH/PN transmissions ($T_2W_2 = 10$ and $I_2W_2 = 10^2$). The two kinds of detection are equalized where the curves cross zero dB.

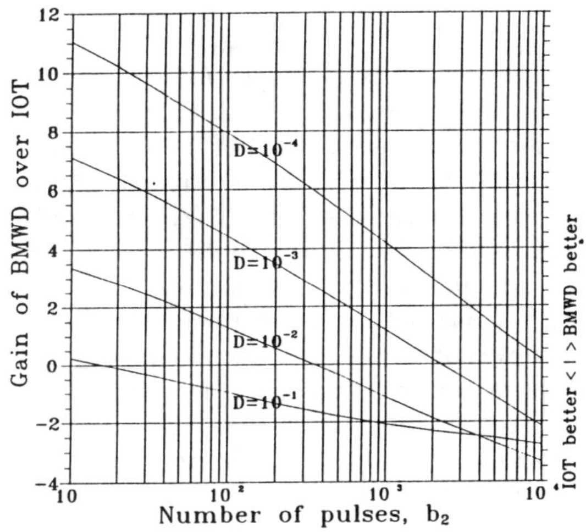

Figure 4.12 The dB difference between the value of E_1/N_{01} required for integration over the transmission and that for pulse detection with BMWD, for TH, FH, and FH/TH transmissions; $T_2W_2 = 1$, $D = b_2/T_1W_1$, $P_d = 0.5$, FAR $= W_1 \times 10^{-11}$

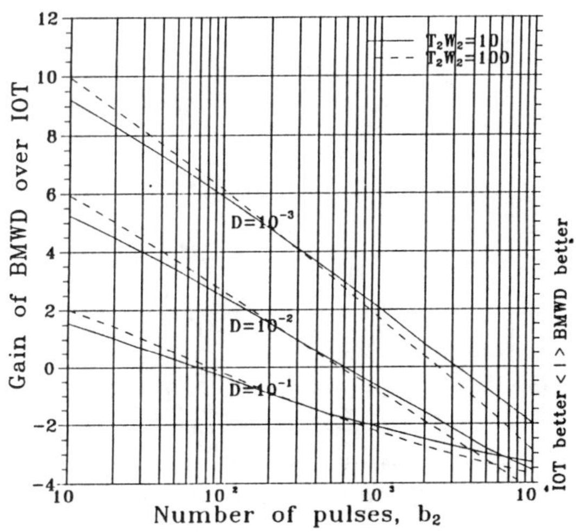

Figure 4.13 The dB difference between the value of E_1/N_{01} required for integration over the transmission and that for pulse detection with BMWD, for TH/PN, FH/PN, and FH/TH/PN transmissions; $D = b_2T_2W_2/T_1W_1$, $P_d = 0.5$, FAR $= W_1 \times 10^{-11}$.

Several potential kinds of loss are disregarded in Figures 4.12 and 4.13. The BMWD cases involving TH assume synchronization with the time-hop interval. Both IOT and pulse detection will suffer a signal-splitting loss (L_t, described in Section 3.3) from integrate-and-dump energy detection. Filter-bank pulse detection may suffer a signal-splitting loss from misalignment of the filters with the hop frequencies.

In Figures 4.12 and 4.13, we note that at zero dB (where energy detection over the transmission and pulse detection with a BMWD are equally effective), the product b_2D ranges from 1 to about 6 or 7. Recalling that $D = 1/N$ for FH and FH/PN (for $N = W_1/W_2$ nonoverlapping filters), we see that b_2 ranges between N and about $7N$ for equalized performance. This is consistent with the rough comparisons described in [2], where it was observed that if the number of filters exceeds the number of pulses, the channelized receiver requires approximately the same power or less than a wideband radiometer. This observation can also be made in Figure 4.14, which was plotted by using the zero crossings from Figures 4.12 and 4.13. Figure 4.14 agrees fairly closely with the original version in [1]. The relationship $b_2D = 1$ is equivalent to $b_2^2 = T_1W_1/T_2W_2$, which would appear as a diagonal line from the lower left corner to the upper right corner in Figure 4.14. Given larger pulse sizes, the lines for that relationship shift to the left, which is roughly consistent with the curves shown.

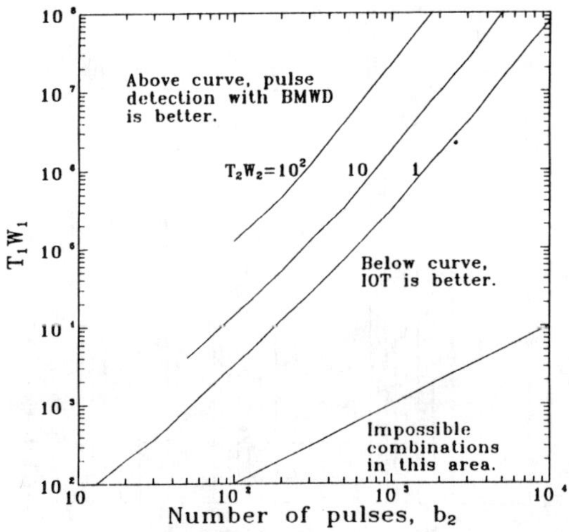

Figure 4.14 For combinations of T_1W_1 and b_2 that fall above the curve corresponding to the selected pulse size, pulse detection using a BMWD is the better detection method. For T_1W_1 and b_2 below the curve, integration over the transmission is the better detection method. ($D = b_2T_2W_2/T_1W_1$, FAR $= W_1 \times 10^{-11}$, $P_d = 0.5$.) *Source:* After Dillard [1], Figure 10.

REFERENCES

1. Dillard, R. A., "Detectability of Spread Spectrum Signals," *IEEE Trans. on AES,* Vol. AES-15, No. 4, July 1979, pp. 526–537.
2. Torrieri, D. J., *Principles of Secure Communication Systems,* Norwood, MA: Artech House, 1985.

Appendix A
Likelihood-Ratio Detection

The optimum detection procedure uses the *likelihood ratio* to decide if a signal is present. The likelihood ratio is defined as follows. A random variable V has probability density function $f(v,\lambda)$, where λ is an unknown parameter. (In the detection problem considered, λ is related to the signal energy.) The decision problem is to choose between hypothesis $H_0{:}\lambda = \lambda_0$ and $H_1{:}\lambda = \lambda_1$. For a single sample v of the random variable V, we compute

$$L = \frac{f(v;\lambda_1)}{f(v;\lambda_0)} \tag{A.1}$$

and compare L with a threshold L_0. If $L \geq L_0$, we decide in favor of H_1; otherwise, we accept H_0. The threshold L_0 is usually chosen so that $P[H_1|H_0]$ has some prescribed value, where $P[H_p|H_q]$ denotes the probability of accepting H_p given that H_q is true. In detection applications, $P[H_1|H_0]$ is the *probability of a false alarm* (PFA), and $P[H_1|H_1]$ is the probability of detection.

When we have N samples $v_1, v_2, \ldots, v_N$ that are *independent and identically distributed* (IID), the likelihood ratio is

$$L = \frac{f(v_1,v_2,\ldots,v_N;\lambda_1)}{f(v_1,v_2,\ldots,v_N;\lambda_0)} = \prod_{k=1}^{N} \frac{f(v_k;\lambda_1)}{f(v_k;\lambda_0)} \tag{A.2}$$

By using logarithms, (A.2) can be expressed as the sum:

$$\log L = \sum_{k=1}^{N} \log \frac{f(v_k;\lambda_1)}{f(v_k;\lambda_0)}, \tag{A.3}$$

which is compared with $\log L_0$.

A case of interest is when we have N independent random variables $v_1, v_2, \ldots, v_N$ and consider the following hypotheses. The hypothesis H_0 is that v_k has density $f(v_k;\lambda_0)$ for $k = 1,2, \ldots, N$; that is, the N random variables are IID. The hypothesis H_1 is that for some j, v_j has density $f(v_j;\lambda_1)$ and the remaining $N - 1$ random variables $v_1, \ldots, v_{j-1}, v_{j+1}, \ldots, V_N$ are IID with density $f(v_k;\lambda_0)$. The index j is a random selection from $\{1,2, \ldots, N\}$ and the probability that j is selected is $P[j]$. For this case, the conditional likelihood ratio, conditional on the index j, is

$$L|j = f(v_j;\lambda_1) \prod_{\substack{k=1 \\ k\neq j}}^{N} f(v_k;\lambda_0) \Big/ \prod_{k=1}^{N} f(v_k;\lambda_0) \tag{A.4}$$
$$= \frac{f(v_j;\lambda_1)}{f(v_j;\lambda_0)}$$

We average (A.4) over the possible values of j to obtain

$$L = \sum_{j=1}^{N} P[j]L|j = \sum_{j=1}^{N} P[j]\frac{f(v_j;\lambda_1)}{f(v_j;\lambda_0)} \tag{A.5}$$

When any value of j is equally likely (i.e., $P[j] = 1/N$), (A.5) reduces to

$$L = \frac{1}{N}\sum_{j=1}^{N} \frac{f(v_j;\lambda_1)}{f(v_j;\lambda_0)} \tag{A.6}$$

Next, suppose we have b sets of N samples of the form just discussed, where a new index j is chosen randomly and independently for each set. We apply (A.2) to (A.5), with an appropriate change of notation, to obtain the likelihood ratio:

$$L = \prod_{i=1}^{b}\sum_{j=1}^{N} P[j]\frac{f(v_{ji};\lambda_1)}{f(v_{ji};\lambda_0)} \tag{A.7}$$

As above, when $P[j] = 1/N$, (A.7) reduces to

$$L = N^{-b}\prod_{i=1}^{b}\sum_{j=1}^{N} \frac{f(v_{ji};\lambda_1)}{f(v_{ji};\lambda_0)} \tag{A.8}$$

Because N and b are known, the factor N^{-b} can be absorbed into the constant L_0. Also, by taking logarithms, we obtain

$$\log L = \sum_{i=1}^{b} \log \sum_{j=1}^{N} \frac{f(v_{ji};\lambda_1)}{f(v_{ji};\lambda_0)} \tag{A.9}$$

Urkowitz [A1] shows that the normalized output $V' = 2V/N_{01}$ from an energy detector has a noncentral chi-square distribution with $2TW$ degrees of freedom and noncentrality parameter $\lambda = 2E_S/N_{01}$, where E_S is the total signal energy and N_{01} is the one-sided noise power spectral density. When $\lambda = 0$ (i.e., when no signal is present)V' has the (central) chi-square distribution with $2TW$ degrees of freedom.

The density of a random variable that has the noncentral chi-square distribution with $2TW$ degrees of freedom and noncentrality parameter λ is [A1]:

$$f(v;\lambda) = \frac{1}{2}\left[\frac{v}{\lambda}\right]^{(TW-1)/2} \exp\left[-\frac{v+\lambda}{2}\right] I_{TW-1}(\sqrt{v\lambda}) \tag{A.10}$$

where $I_\gamma(x)$ is the modified Bessel function of the first kind, of order γ. If we expand the Bessel function in a power series and evaluate (A.10) when $\lambda = 0$, we obtain the density:

$$f(v;0) = \frac{v^{TW-1}e^{-v/2}}{2^{TW}\Gamma(TW)} \tag{A.11}$$

We substitute (A.10) and (A.11) into (A.1) to obtain the likelihood ratio:

$$L = \left[\frac{2^{TW-1}\Gamma(TW)e^{-\lambda/2}}{\lambda^{(TW-1)/2}}\right] \frac{I_{TW-1}(\sqrt{v\lambda})}{v^{(TW-1)/2}} \tag{A.12}$$

Because the bracketed term in (A.12) does not depend on v, it can be absorbed into the constant L_0.

Equation (A.5) applies to frequency hopping (FH) when N is the number of filters. Equation (A.6) also applies to FH if exactly one filter contains signal at each energy-detection time. Equation (A.5) applies to time hopping (TH) if the energy detector is synchronized with the hopping intervals and we reinterpret the sum to be over time. If the signal is equally likely to appear in any time slot within the hopping interval, (A.6) also applies to TH. For $TW = 1$ (i.e., for pure FH or pure TH), we compute

$$L = \sum_{j=1}^{N} I_0(\sqrt{v_j\lambda}) \tag{A.13}$$

and for $TW > 1$ (i.e., for FH/PN or TH/PN), we compute

$$L = \sum_{j=1}^{N} \frac{I_{TW-1}(\sqrt{v_j \lambda})}{v^{(TW-1)/2}} \tag{A.14}$$

If we interpret N to be the product of the number of filters and the number of time slots per hop interval, (A.13) applies to FH/TH and (A.14) applies to FH/TH/PN.

Equations (A.10) through (A.14) are in terms of the normalized energy detector output $V' = 2V/N_{01}$. To represent the unnormalized output V, we make the change of variable $v_j = 2z_j/N_{01}$ in (A.13) and (A.14). We also make the change of variable $\lambda = 2E_S/N_{01}$ to obtain

$$L = \sum_{j=1}^{N} I_0\left(\frac{2}{N_{01}} \sqrt{z_j E_S}\right) \tag{A.15}$$

and

$$L = \sum_{j=1}^{N} \frac{I_{TW-1}\left(\frac{2}{N_{01}} \sqrt{z_j E_S}\right)}{\left(\frac{2}{N_{01}} z_j\right)^{(TW-1)/2}} \tag{A.16}$$

for the $TW = 1$ and $TW > 1$ cases, respectively.

REFERENCE

A1. Urkowitz, H., "Energy Detection of Unknown Deterministic Signals," *Proc. IEEE*, Vol. 55, No. 4, April 1967, pp. 523–531.

Glossary

Symbol or Acronym	Definition
BMWD	Binary moving window detector
BPF	Bandpass filter
b	Number of pulses or data symbols
b_2	Number of data symbols in a transmission
b_3	Number of elements (pulses) in a data symbol
D	Density; ratio of the time-bandwidth area occupied by signal to the area of the transmission ($D = b_2T_2W_2/T_1W_1$)
E, E_s	Signal energy
E_1	Signal energy of transmission
E_2	Signal energy of data symbol
E_3	Signal energy of element in a data symbol
F	Q_F/TW; a constant factor of FAR
FAR	False-alarm rate (alarms/s)
FB-ED-OPT	Filter-bank energy-detection optimum (likelihood ratio) system
FB-ED-OR/BMW	Filter-bank energy-detection OR-gate binary-moving-window system
FB-ED-T	Filter-bank energy-detection thresholding system
F-ED-BMW	Single-filter energy-detection binary-moving-window system
F-ED-C/T	Single-filter energy-detection time-collapsing thresholding system
F-ED-OPT	Single-filter energy-detection optimum (likelihood ratio) system
F-ED-OR/BMW	Single-filter energy-detection OR-gate binary-moving-window system
F-ED-T	Single-filter energy-detection thresholding system
FH	Frequency hopping

Symbol or Acronym	Definition
G_{MW}	Gain of a system using binary moving-window detection over a system using simple thresholding
I+D	Integrate-and-dump energy detection
K	Energy detector threshold
K_M	Threshold of the BMWD
K_0	Normalized energy detector threshold ($K_0 = K/2$)
L	Likelihood ratio statistic
L_f	Loss due to splitting signal in frequency (into two filter bands)
L_t	Loss due to splitting signal in time (using I + D detection)
L_{t+f}	Loss due to splitting signal in frequency and time
L_r	Loss of radiometer compared to ideal energy detection; known as radiometer constant or degradation factor
M	Length (in bits) of a binary moving window
M_2	Number of integration cells (in a time-bandwidth grid) containing only noise
N	Number of filters
N_{01}	One-sided noise power spectral density
n	Number of (radar) pulses equivalent to energy detection area
PC	Pulse compression
P_d	Probability of detecting transmission
PN	Pseudonoise
P_T	False-alarm probability of the BMWD
p_s	Probability of coinciding with and detecting any particular pulse when frequency scanning
p_0	Probability of a zero from the binary moving window
p_1	Probability of a 1 from the binary moving window
Q_D	Energy detector detection probability
Q_F	Energy detector false-alarm probability
q	Frequency-scanning ratio W_2/sT_2
r	TH: $r = T_{p2}/T_2$; frequency scanning: $r = W/(W_s + sT_2)$
S	Signal-energy to noise-density ratio; equivalent to E_s/N_{01}
S-ED-BMW	Frequency-scanning energy-detection binary-moving-window system
S-ED-T	Frequency-scanning energy-detection thresholding system
S/PC-ED-BMW	Frequency-scanning pulse-compression energy-detection binary-moving-window system
S/PC-ED-T	Frequency-scanning pulse-compression energy-detection thresholding system
s	Slope (frequency/time) of frequency-scanning sweep
T	Integration time of energy detector

Symbol or Acronym	*Definition*
T_C	Integration time of energy detector after pulse compression
TH	Time hopping (hopped)
T_M	Binary integration time of a moving window detector
T_{p2}	Duration of time-hopping interval
T_{p3}	Duration of time-hopping interval for elements within data symbol
T_1	Duration of transmission
T_2	Duration of data symbol
T_3	Duration of element of data symbol
t_d	Duration of look in each direction per scan in azimuth
t_s	Duration of scan over frequency or azimuth
W	Bandwidth of energy integration
W_s	Surveillance bandwidth of frequency-scanning system
W_0	$W_s + W$, total sweep of frequency-scanning system
W_1	Bandwidth of transmission
W_2	Bandwidth of data symbol
W_3	Bandwidth of element (of data symbol)
λ	Noncentrality parameter $2E_s/N_{01}$
ρ	Single-pulse/signal-to-noise ratio (SNR)
θ_d	Beamwidth, in azimuth, of antenna scanning
θ_s	Angle of angular sector of azimuth scanning

Index